A Topological Picturebook

George K. Francis

A Topological Picturebook

With 87 Illustrations

Springer-Verlag
New York Berlin Heidelberg
London Paris Tokyo

George K. Francis
Department of Mathematics
University of Illinois at
  Urbana-Champaign
Urbana, IL 61801

The cover illustration is a
photograph of a drawing on a blackboard.

AMS Classification: 54–01, 51H05

Library of Congress Cataloging in Publication Data
Francis, George K.
  A topological picturebook.
  Bibliography: p.
  Includes index.
  1. Topology—Graphic methods. I. Title.
QA611.F69   1987   514   86-31400

© 1987 by Springer-Verlag New York Inc.
All rights reserved. This work may not be translated or copied in whole or in part without the written permission of the publisher (Springer-Verlag, 175 Fifth Avenue, New York, New York 10010, USA), except for brief excerpts in connection with reviews or scholarly analysis. Use in connection with any form of information storage and retrieval, electronic adapation, computer software, or by similar or dissimilar methodology now known or hereafter developed is forbidden.
The use of general descriptive names, trade names, trademarks, etc. in this publication, even if the former are not especially identified, is not to be taken as a sign that such names, as understood by Trade Marks and Merchandise Marks Act, may accordingly be used freely by anyone.

Printed and bound by R.R. Donnelley & Sons, Harrisonburg, Virginia.
Printed in the United States of America.

9 8 7 6 5 4 3 2 1

ISBN 0-387-96426-6 Springer-Verlag New York Berlin Heidelberg
ISBN 3-540-96426-6 Springer-Verlag Berlin Heidelberg New York

To Bettina

PREFACE

Die Kräftigung des räumlichen Vorstellungsvermögens und der räumlichen Gestaltungskraft gehört unbestritten zu den wichtigsten Zielen eines jeden geometrischen Unterrichts.
　　　　　　　Artur Schoenflies, 1908

Surely, among the most important goals of every geometrical instruction is the strengthening of the faculty for spatial imaging and the power for spatial modelling.

My book is about how to draw mathematical pictures. Of course, it begs the question of whether one should draw pictures in mathematics in the first place. Some disciplines require illustrations. Imagine Gray's *Anatomy* without pictures. Some get along better without them. Imagine trying to illustrate Russell and Whitehead's *Principia*.

The decade of this book's gestation began with a sabbatical semester I spent in our library with the master geometers of the previous century. In the pages of the *Enzyclopaedie der Mathematischen Wissenschaften* and the collected works of Felix Klein I began to see, through hundred-year old eyes, what I had learned as a mathematics student in the middle of this century.

Theirs was a wonderfully straightforward way of looking at rather complicated things, notably Riemann surfaces and geometrical constructions over the complex numbers. They drew pictures, built models and wrote manuals on how to do this. And so they also captured a vivid record of the mathematics of their day. I resolved to try to do the same for the mathematics of my contemporaries.

In my own work I had always relied heavily on sketches and diagrams. I learned really to appreciate this experimental way of doing mathematics during my Ann Arbor days. There Chuck Titus guided me skillfully through my thesis by drawing hundreds of pictures. I also adopted his historical approach towards geometrical analysis, which leads to a harmony between logically abstract and visually concrete mathematics. Encouraged by the leaders in my field, Morris Marx in particular, I entertained my lecture and seminar audiences with elaborate blackboard drawings and pictorial "souvenir" handouts. Eventually, I began to work on pictures of things other people wanted to have portrayed. I had the pleasure of illustrating an article by Berstein and Edmonds [1979], a Springer monograph by Bill Abikoff [1980] and his essay in the *Monthly* [1981], two papers by Bill Thurston [1982,1986] and his as-yet-unpublished textbook, *The Geometry and Topology of Three-Manifolds*. And, above all, I now actively badgered my colleagues to tell me their secret visions with the promise of showing them a simple way of sketching them on demand.

When I first decided to formalize the procedures I had developed for drawing useful mathematical pictures, I planned on calling the result "A Graphical Calculus." I still believe that there are some rules, based on differential geometry, which can be distilled into practical routines for "calculating" how to draw a picture. In the meantime, however, I have come to favor a less didactic, more advisory purpose for setting down any rules. You should regard this book as a description of my own graphical calculus and as an invitation for you to modify it and improve upon it to suit your own purposes.

The pictures in this book are certainly meant to be copied or reproduced, peferably by hand. They also can be traced onto a transparency. Most pen drawings "work" equally well without the shading I have applied to them for effect. Each picture has a "line pattern" from which it was constructed. In many instances I included this pattern explicitly. In others you may rediscover it by analysing the finished picture. The line pattern is usually simple enough to commit to memory or to your lecture notes. The pictures have been tested out at the blackboard to make sure that this mode of presentation is also feasible. Even the more elaborate ones, using colored chalk, can be reproduced in half an hour. Yet, the pedagogical advantage of composing a picture at the board, erasing the hidden lines, highlighting the curvature and using colored chalk to label details is compromised by the impermanence of this time-consuming medium. The only efficient way to use it for talks is to make photographic slides of the drawings. Therefore the book concentrates on how to make india-ink line drawings that are robust enough to survive several xerographic generations and simple enough to reproduce by hand.

A word about the format I chose and how this topological picture-storybook got its name. I still have the *Märchen-Bilderbuch* my grandfather gave me for my third birthday. He had bound it himself. My oldest possession, this beautifully illustrated volume of Grimms' tales [1942], survived wartime bombs and postwar turmoil. I have always loved this literary genre. So when my patient and generous Springer editor, Walter Kaufmann-Bühler, suggested I add more text by way of explanation, it seemed natural to tell topological stories to go with the pictures.

This way each chapter became a "picture story" about something in a particular area of mathematics. Each owes its telling to expert friends who one way or another told me about their territory. Serendipity selected the subjects that came my way. But responsibility for the text, whether clear or not, as well as for the drawings, whether pleasing or not, is strictly my own.

Chapter 1 establishes the relationship between Whitney and Thom's theory of general position, of stable singularities of low-dimensional mappings, and descriptive topology. The drawings were originally prepared for Robin McLeod's Workshop on Frontiers of Applied Geometry at Las Cruces, New Mexico in 1980. I have given this lecture many times since and directed it especially to those who draw pictures to go along with their own lectures.

The techniques in this and the next chapter were originally collected for my efforts in using pictures to explain what Catastrophe Theory is all about. By such visual means I was able to "explain" Thom and Zeeman's intriguing invention also to non-topologists, in the sense that I was able to arouse and satisfy their curiosity. My courses for high-school math teachers in Peoria and biology students at the University and my seminars for physics and physiology colleagues all relied on the *lingua franca* of accurately drawn pictures. I found these pictures in the literature and learned how to reproduce them from the drawing of Jim Callahan, Tim Poston and Les Lander in Callahan [1974,1977], Poston-Stewart [1978], and Bröcker-Lander [1975] respectively.

Chapter 2 is the *media and methods* section of the project, a form I borrowed from the laboratory sciences. Topologists usually draw pictures to help formulate their definitions and proofs. On the way to publication, these visual props are all too often left behind. Here then is a fair sample of the various ways a topologist may produce pictures suitable for publication. This chapter is more of a collection of topological "anecdotes" than a coherent picture story. Most of them were originally prepared for Bill Abikoff's long lived and much-appreciated Riemann Seminar, where we studied the work of Thurston *inter alia*. The last five figures in this chapter were "commissioned" executions of Thurston's sketches.

I am often asked whether my pictures are drawn by a computer. Not yet! My efforts first to learn and then to enlist this difficult medium into the service of topology began as long ago as the picturebook project. How I used computer output for my drawings is explained herein. I wish to thank my tutors, Judy and Bruce Sherwood, who initiated me into the electronic mysteries of PLATO. Only recently has it been possible for me to use a computer with anywhere near the ease and facility of a pen. I have included a few color photographs of computer-drawn surfaces designed in a manner outlined in this book. They were made by Donna Cox and Ray Idaszak on machines provided by the Electronic Imaging Lab and the National Center for Supercomputing Applications at the University of Illinois. My postscript, of a somewhat technical nature, tells the story of these pictures.

Chapter 3 is a review of the principles of perspective drawing from the viewpoint of projective geometry. Its terse and didactic style is borrowed from descriptive geometry. It centers on three elementary problems a draftsman faces when drawing in freehand perspective. I learned about these and their various solutions from Norm MacFarland through the University of Illinois Program for Faculty Study of a Second Discipline. The semester I was a guest in Ed Zagorski's Department of Industrial Design introduced me to an artistic discipline with remarkably strong affinity to descriptive topology. Invaluable drafting lessons from Vivian Faulkner-King led to my redesigning some and redrawing most of the older pictures.

For completeness I added a section on non-perspective, so called "engineering" drawing, which — unfortunately — is still used by many publishers of mathematical textbooks. This style, with its illogical distortions, was originally developed to convey metric information from designer to machinist. Computer aided design does this much better nowadays, and computer graphics makes perspective views trivially accessible to all. With some reluctance I also included a section on shading. Despite many hours of careful inspection of Escher, Dürer, and Dali's works, I don't think I have mastered the technique of light and shade well enough to teach it.

In this chapter more than in the others I have tried to speak the language of the artist and designer. While the main purpose of this book is to persuade fellow topologists to draw the pictures that illustrate their work, I hope it will also inspire artists to study topology and apply their skill to this wonderfully visual discipline.

Chapter 4 is a short story about the optical illusion on which several of Maurits Escher's best known prints are based. Here it serves as an introduction to the differential geometry of spaces of constant curvature. Originally I distributed the drawings as a geometric curiosity. At times I thought the impossible tribar had a natural home in a lens space. For the true story, I am indebted to many colleagues who spotted such errors and suggested improvements, notably Wolfgang Haken, Allan Hatcher, Dave Fried and John Stillwell.

Chapter 5 is algebraic geometry in a most elementary and practical sense: just what should a picture of an algebraic operation, such as squaring or multiplying, really look like? Building on common knowledge of ninth-grade analytic geometry, which gives answers over the field of real numbers, I draw the shadows of four-dimensional shapes that give answers over the complex number field. My debt to the classic computer graphics films by Tom Banchoff and Charles Strauss [1977] is clear. This chapter was particularly influenced by their treatment of the complex function graphs and the Veronese Surface. However, the pedagogy that underlies the entire book, and which I bring out specially here, comes from Bernard Morin of Strasbourg. Pictures without formulas mislead, formulas without pictures confuse. I don't know if Bernard would say it this way, but it is how I have understood his work

The chapter concludes with applications of the graphics and of the mathematics. In the former I add some pictures to the rarely referenced, highly visual part of Whitney's famous treatise [1941], where he gives a concrete meaning to his theory of characteristic classes. The latter refers to the splendid thesis by the Morin student, François Apery [1984]. It constitutes a gateway between traditional hand-drawn graphics and efficient computer graphics, a subject to which I return in the postscript.

Chapter 6 is, in a sense, the first picture story. It relates once more, and with a different emphasis, the details of a famous visualization problem in differential topology. When I tried to explain the sphere eversion to my freshman topology seminar in 1977 I already had the *Scientific American* cover article by Tony Phillips [1966] and the NSF-sponsored computer graphics film by Nelson Max [1977] at my disposal. Still, it took much effort and many blackboard sessions to show my students what they had, in fact, already seen. In time it became clear that the few ink drawings that fit on a single page could suffice to demonstrate an eversion; see pages 124 and 113. The design of these pictures was based on a well-established body of theorems on the classification of excellent mappings, Francis-Troyer [1977,1982]. Thus the demonstration of the tobacco pouch eversion could be made as rigorous as one's taste dictated. In the end, however, it was Bernard himself who provided the analytic parametrization for this remarkable deformation of a surface; see Morin [1978]. I hope that you too will recognize the intimate relation of topology, art and analytic geometry, at least in this singular case.

Chapter 7 began as an orphan. In the course of studying Hatcher and Thurston's proposed scheme for presenting the mapping class group of an arbitrary surface I had drawn some pictures for our Riemann Seminar. They illustrated details in the notes by André Marin [1977], which at that time were circulating in the mathematical *samizdat*. Already in the first manuscript circulated by the authors, and decidedly in their final version [1980], a revision of their argument made my pictures largely superfluous. I decided I would instead apply my efforts to illustrating the topology behind the well-known presentation of the mapping class group of the double torus. Combinatorial group theory is a field the furthest from my talents that I ever dared approach. That I succeeded in my quest is to the credit of my mentors, Bill Abikoff, Joan Birman, John Ratcliffe, Paul Schupp, and especially Wilhelm Magnus, who provided endless hours of patient explanation to go with a generous gift of his collected works. I hope that the pictures in this chapter are only the beginning of an endeavor, joined by others, to enrich the visual side of this abstract but beautiful mathematical discipline.

Chapter 8 is about a fibered knot and so belongs to geometric topology. Since the knot is the star of Thurston's book [1977,1982], my story belongs to the larger domain of low-dimensional geometry and topology. At any rate, here it is again together with drawings and side trips that I left out of its original version, published in the *Monthly* [1983]. That article was a trial balloon for an expository style suitable for the picturebook. The excellent suggestions of its editor, Paul Halmos, its four referees (two known to me, two not), and the kind comments of its readers eventually fused into the genre I call a *topological picture story*.

I don't think the picturebook manuscript would ever have been finished without the helpful criticism of friendly readers and the persistence of my publisher. It would have remained an embarrassment of obscurity without the veritable flood of red ink, for which I am eternally grateful to Ian Stewart. The project has also older debts. Early and awkward versions of most ideas herein were foisted on and loyally dissected by fellow members of the Geometric Potpourri Seminar: Ralph and Stephanie Alexander, Felix Albrecht, Dick Bishop and Dave Berg. Hermann Rusius of Enschede, Holland provided a rare sort of encouragement which deserves special mention. Cloth fabrics merchant by profession, self-taught geometer and builder of mathematical models, in the winter of 1979 he arrived in Urbana in a blizzard and left again with the next. In between he photographed the dusty plaster models of algebraic surfaces made by Verlag Schilling of Leipzig in the last century that now lie entombed in the glass cases of Altgeld Hall. We spent many evenings together talking shop. Later, when it came time to write about my pictures and I was at a loss to think of my proper audience I secretly imagined Mr. Rusius and tried to explain it to him.

And of course, no project of this absorbing a nature could have been sustained over such a lenghty span of time without long-suffering forebearance and cheerful support from my wife and children.

At critical moments in the production of this book I received expert help from many old and new friends in the book printing profession. Special thanks go to editors Rob Torop and Barbara Tompkins of Springer-Verlag, to Carole Appel, senior editor at the University of Illinois Press, and to Larry Lutz, Carl Kibler and their colleagues at the University of Illinois Print Shop, where the glow of electronic typography reflects off venerable Monotype machines, and the smell of photographic chemicals mixes with the nostalgic odor of hot lead.

Urbana, Illinois
September 1986

George K. Francis

CONTENTS

Reference to a detail of a figure in the current chapter is by number, row and column. If the figure is located in another chapter then its number preceeds the cross reference. Thus the Möbius band in the lower right corner of Figure 9 in Chapter 6 is called 6:9(34). Chapter subheadings which are also figure captions are preceded by the figure number and followed by their page number.

COLOR PLATES (following page 16)
- I BLACKBOARD DUNCE CAP
- II DIAPERED TREFOIL KNOT
- III CROSS CAP AND LIMAÇON
- IV ETRUSCAN VENUS

CHAPTER 1 DESCRIPTIVE TOPOLOGY ... 1
1. SADDLE IN A BOX ... 2
2. SADDLE IN A DRUM ... 4
3. WHITNEY UMBRELLA ... 7
4. CAYLEY CUSP ... 10
5. PINCH POINT/BRANCH POINT ... 12

CHAPTER 2 METHODS AND MEDIA ... 14
1. CUBIC METAPHOR ... 15
2. BASIC SURFACE PATTERNS ... 17
3. TRIPRONG ... 18
4. ROUNDING A POLYHEDRAL SADDLE ... 19
5. DUNCE HAT ... 21
6. DUNS EGG ... 23
7. MÖBIUS BANDS ... 25
   - INK AND PAPER ... 24
   - CHALK AND BLACKBOARD ... 26
   - SLIDE AND TRANSPARENCY ... 28
   - COMPUTER AND DRAFTING TABLE ... 29
8. FUNCTION GRAPHS ... 30

9	Cancelling Pinch Points: Wire Frames	32
10	Cancelling Pinch Points: Hidden Lines	33
11	Cancelling Pinch Points: Elaborations	34
12	Spinning a Pair of Pants	35
13	Lamination and Eight Knot	37
14	Cable Knot and Companion	38
15	Cabling Template	39
16	Octahedral Hyperbolic Manifold	40
17	Tetrahedral Hyperbolic Manifold	41
18	Borromean Orbifold	42

Chapter 3 PICTURES IN PERSPECTIVE — 43

1	Linear Perspective	44
2	Horizon and Zenith	47
3	Recipe for a Cube	49
4	Cube in 3-2-1 Point Perspective	51
5	Looking for a Cube	53
6	Diagonals	55
7	Wheel and Axle	57
8	Axonometrics	58
9	Chiaroscuro	61

Chapter 4 THE IMPOSSIBLE TRIBAR — 65

1	Penrose Tribar	66
2	Impossible Quadrilateral	67
3	Pipeline	69
4	Dice	70
5	Poincaré Group	71
6	Cubic Lattice	73
7	Man in a Cube	74
8	Left Hand/Right Hand	76

Chapter 5 SHADOWS FROM HIGHER DIMENSION — 77

1	Crossing a Channel	79
2	Plücker Conoid	81
3	Pinching a Roman Surface	84
4	Gauss Map and Cross Cap	87
5	Boy Surface	90
6	Slice and Shadow	93
7	Whitney Bottle	94
8	Romboy Deformation	95

Chapter 6 SPHERE EVERSIONS — 99

1	Golden Rectangle	100
2	Cross Cap and Handle	101
3	Forbidden Eversion	103

	4	BASEBALL MOVE	105
	5	CHAPEAU DIAGRAM	108
	6	ASTROID/DELTOID	109
	7	ASSEMBLING IMMERSIONS	111
	8	MORIN EVERSION	113
	9	TOBACCO POUCHES	115
	10	ASTROID FAMILY	118
	11	CATASTROPHE MACHINE	121
	12	MORIN TWIST	123
	13	ORIGINAL DESIGN	124

CHAPTER 7 GROUP PICTURES — 125

	1	ZIPPING UP A DOUBLE TORUS	127
		PRESENTING THE DOUBLE TORUS	129
	2	BRAIDING HOMEOMORPHISM	130
	3	ARTIN SWAP	132
	4	SPHERICAL BRAIDS	133
	5	PLATE TRICK	135
	6	WHORL AND CHIMNEY	137
	7	SWAPPING HANDLE CORES	140
	8	SWAP DIAGRAM	141
	9	LICKORISH TWIST	142
	10	KING SOLOMON SEAL	145

CHAPTER 8 THE FIGURE EIGHT KNOT — 149

	1	PROJECTIONS OF THE KNOT	150
	2	HEXAHEDRAL COMPLEMENT	152
	3	HEXAHEDRAL GLUING DIAGRAM	154
	4	SEIFERT SPANNING SURFACE	156
	5	SIX HAKEN SURFACES	158
	6	FIBER MNEMONIC	159
	7	THE OWL AND THE PUSSYCAT	161
	8	ISOTOPIC SEIFERT SURFACES	163
	9	TRIVIAL FIBRATION	164
	10	DOUBLE TORUS KNOT	165
	11	TETRAHEDRAL EIGHT KNOT: ONE	167
	12	TETRAHEDRAL EIGHT KNOT: TWO	168
	13	TOROIDAL SHEAR ISOTOPY	170
	14	CYLINDRICAL SHEAR ISOTOPY	171
	15	COMPUTING THE MONODROMY	172
	16	HOPF FIBRATION	175

POSTSCRIPT — 176

BIBLIOGRAPHY — 180

INDEX — 186

1
DESCRIPTIVE TOPOLOGY

Descriptive topology, namesake of *descriptive geometry*, serves contemporary topology in the same manner as its forbearer served the geometry of the nineteenth century. In the eleventh edition of the Encyclopedia Britannica, Henrici defines descriptive geometry as that branch of the discipline "which is concerned with the methods of representing solids and other figures in three dimensions by drawings in one plane." The rigor with which this subject used to be taught was justified by its application in the preparation of accurate architectural and engineering drawings. Photography first, and more recently, computer aided design, have greatly reduced the need for its study in our schools. At the same time, the immense expansion of topology in this century, both as a pure and as an applicable science, has produced a sizeable collection of basic examples and a variety of styles for their exposition. "Descriptive topology" is an apt name for their systematic study.

It is a disservice to the unity of mathematics to isolate these examples from their analytic origins and present them as so much "rubber sheet geometry." Since the days of Descartes, expressing geometrical information in that universal language of mathematics, algebra, has been immensely useful in the service of precision and economy of thought. Nevertheless, something is inevitably lost in this transcription. The task of descriptive topology is to unfold the visual secrets so often compressed into algebraic shorthand. To illustrate how this might be done sytematically, I shall draw on Whitney and Thom's theory concerning the generic forms of a surface extended in space or mapped into a plane. The graphical elements of my method are pictures of the canonical algebraic forms associated with surfaces in *general position*. These are *stable* in the sense that slight perturbations of their analytical formulation, or their physical expression, as on a computer screen, do not change them qualitatively. You will find a comprehensive introduction of the mathematics of stable mappings and their singularities in the text by Golubitsky and Guillemin [1973].

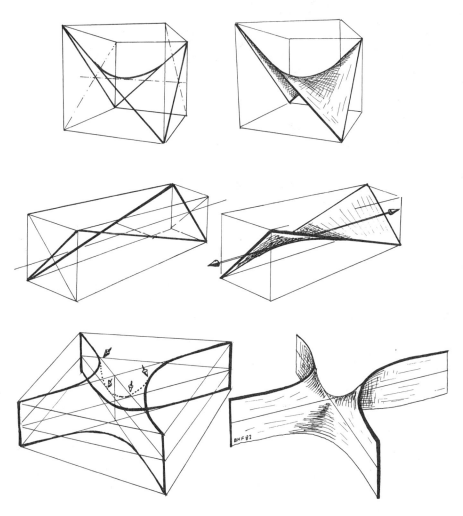

SADDLE IN A BOX

Figure 1

Let us start with the problem of graphing the *hyperbolic paraboloid* whose equation is $Z - XY = 0$. This saddle-shaped surface regularly taxes the artistic abilities of calculus teachers. Here are several ways to approach this problem. First, rewrite the equation to suggest a more imperative mood: *Let $Z = XY$*. This assigns to the Z-coordinate of a point the product of its XY-coordinates. Now treat this product asymmetrically by regarding the first factor as the parameter of a smooth transition between two skew lines in space. In other words, imagine the line $Z = tY$ in the plane $X = t$ turning about the X-axis as the plane moves parallel to itself in the X-direction. Figures 1(11) and 1(12) show such a saddle suspended inside a cube. Stretch the X-scale relative to the other two and you produce a ribbon with a quarter twist as in 1(21) and 1(22). The abbreviation n(ij) refers to the detail in row i column j of figure n in the current chapter. If the figure referred to is in a different chapter, its number comes first. Thus the Möbius band in the lower right corner of Figure 9 in Chapter 6 is called 6:9(34).

A surface which can be regarded as the set of successive position of a curve moving in space is said to be *generated* by the curve. The utility of this notion in constructing a surface geometrically, in a picture or as a model is increased as the complexity of the generator and its motion is decreased. When the generator is a straight line, it is called a *ruled surface*. Since you can exchange X and Y in the above analysis, the hyperbolic paraboloid is generated by a line in two ways. It is a *doubly ruled surface*.

Reverse the equation like this: $XY = Z$ and you say that the point (X, Y) is constrained to lie on a hyperbola whose shape is controlled by the current value of Z. As Z moves from -1 to $+1$, the two branches of the hyperbola draw close to each other, touch in the shape of two crossing lines, and part again, fitting inside the asymptotes "the other way", see 1(31). Freezing this motion in space-time again generates a saddle similar to 1(32). Note the four places (arrows) where the contour of the object in the picture plane is tangent to curves on the surface.

SADDLE IN A DRUM. *Figure 2.*

Change to polar coordinates for another way of visualizing the hyperbolic paraboloid:

$$X = r \cos(t)$$
$$Y = r \sin(t)$$
$$Z = \tfrac{1}{2} r^2 \sin(2t) = a(t) \, r^2 = b(r) \sin(2t).$$

This defines two ways of generating the saddle depending on the order in which the r and t loops are nested. If t is taken for the parameter, the coefficient a of r^2 changes sinusoidally from -1 to $+1$ as t goes from 0 to π. Hence a vertical parabola changes its curvature from down to up as it makes a half turn about its axis of symmetry. Note that in the provisional line pattern, 2(11), there is an undesirable triple point. Can you detect how this coincidence was avoided when I traced 2(12) from this pattern?

If r is taken for the parameter then the scale coefficient b for a bent hoop on a cylinder, 2(21), goes to zero parabolically. The arrows point to important tangencies. The contour is marked as a dotted line. As the hoop shrinks towards the origin, it sweeps out the surface 2(22).

The mode in which analytical expressions and coordinate equations are formulated has considerable influence on the speed and precision with which the reader glimpses the same thing the writer means to describe. At times,

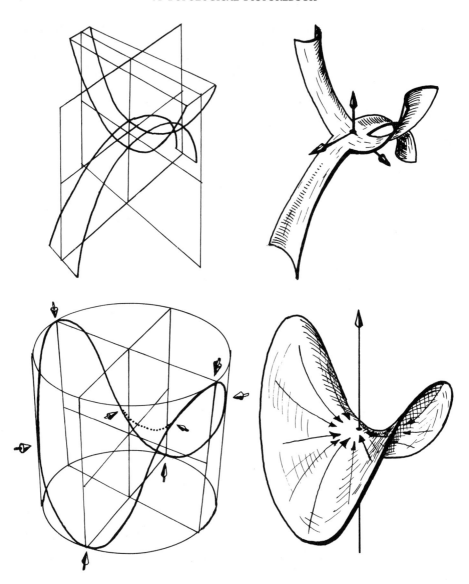

Figure 2 SADDLE IN A DRUM

efficiency requires a departure from customary style in analytical geometry. This is especially true for 3-dimensional objects and phenomena. Take, for example, the locus of points satisfying the polynomial equation $XY - Z = 0$. Unless your reader immediately recognizes the surface from this equation, or proceeds to rewrite the equation, he will probably imagine the following polygonal path: Go from the origin of the coordinate system to a point a (signed) distance X along the first axis, then follow a segment parallel to the second axis, then one in space parallel to the third axis. Do this so as to reach a point whose coordinates fit the constraint.

The visualization problem is reduced by one dimension just by separating the variables thus $XY = Z$. This suggests a family of plane hyperbolas stacked on top of each other to produce the surface. The trick of splitting the dimension of the ambient space becomes essential in describing 4-dimensional phenomena, as we shall see later, especially in Chapter 5.

A further simplification is achieved by turning the equation around: $Z = XY$. This establishes the surface as the graph of a numerical function from the plane. It is also the nontrivial part of a parametrization of the surface. A set of parametric equations is considerably more convenient for drawing a surface than its constraint equation. This is especially true if the drawing instrument is a computer. While it might be feasible to examine each *pixel* (picture element) on a 2-dimensional grid to see whether it satisfies the constraint or not, the 3-dimensional analog requires heroic speed and memory or very sophisticated programming.

There is another problem with the traditional view of a 3-dimensional Cartesian coordinate system. This has to do with our personal experience of space, which is innocent of solids pierced in three directions by ever so fine rods, or finely diced by three sets of parallel planes. There is no difficulty in imagining a 2-dimensional figure drawn on graph paper, to be observed from above like a geographical map. In space, we are more apt to orient ourselves in reference to outside boundaries, the walls of a room for example, than by tracing our position along a right-angled polygon from a corner of the room. In descriptive topology it is, thus, much better to think of coordinates as functions measuring the displacement of points from the faces of familiar reference surfaces (cubes, cyclinders, spheres) encompassing the object than as addresses in a 3-way grid.

Algebraic manipulation of formulas generally provides the transition from a "vectorial" to a "positional" viewpoint. Here are two examples of this. The identity $A + t(B - A) = A(1 - t) + tB$ expresses the "t-th point" on a line segment as being reached from A by moving along the vector $V = B - A$, and as the position which is the fraction t of the way from A to B. The second example concerns alternate ways of expressing a 2-dimensional rotation

$$x \begin{vmatrix} \cos(t) \\ \sin(t) \end{vmatrix} + y \begin{vmatrix} -\sin(t) \\ \cos(t) \end{vmatrix} = \begin{vmatrix} x \\ y \end{vmatrix} \cos(t) + \sin(t) \begin{vmatrix} -y \\ x \end{vmatrix}$$

The left side says that the new position of the point, formerly at *(x,y)*, may be found in the traditional way using a coordinate frame which has been rotated by an angle of t. The second says to move the point *(x,y)* an angle t in the direction of the left-hand perpendicular *(− y,x)* along the circular arc joining these two points, and centered at the origin.

WHITNEY UMBRELLA. *Figure 3*.

Good design of a topological picture involves imagining something in 3-space that embodies the mathematical idea to be illustrated. Then you must draw it in such a way that the viewer has no difficulty in recognizing the idea. The picture should cause him to imagine the same object without his having to consult a long verbal description.

The fabric of the objects I have in mind consists of surface sheets joined to each other along three kinds of curves. Some curves lie entirely in the picture and close upon themselves or end at special points. Others are supposed to extend beyond the borders of the drawing. The problem of designing bounded drawings of unbounded objects arises also in art. The artistic solution of continuing right up to the picture frame, however, does not work very well in topology. Here it is better to draw a piece of the object as it would appear if it were isolated from the rest by a transparent container. You may think of the piece contained in a topological neighborhood of a central point on it. Some prefer ragged edges on a surface piece to suggest that it continues. Besides being harder to draw, such edges are often too busy and draw too much attention to themselves. I prefer to use boxes, spheres or cylinders for the containing neighborhoods.

Ultimately, whether drawn by hand or computer, a picture consists of (curved or straight) lines joining points in a plane, call it the *picture plane*. Suppose P is a point on the picture plane corresponding to the point P' on the depicted object, and D is a small disc in the picture plane centered on P. If D looks like the image of a disc-like patch D' on the object then P is said to be a *regular point*. More precisely, you should be able to regard D' as the graph of a smooth (differentiable) function, $u = f(x,y)$, where x,y are local coordinates on D and u measures the distance from the picture plane to the object. Evidently, all other points on the interior of D are regular also and it is proper to speak of a *regular patch* D' on the object corresponding to its image D in the picture.

For complicated objects it is often impossible to find a view which does not hide some important structure behind a surface sheet. One remedy is to remove a regular patch from the object, creating a transparent *window* through which this structure can be seen in the picture. The reason why you should use a regular patch is that here you can readily infer the correspondence between curves drawn on D and curves on D'. Of course, unless the coordinates and function $f(x,y)$ on D are explicitly given, there is no *a priori* way to infer the exact position of D' in space. The context of the picture must make that clear.

Let me now describe in some detail the kind of surfaces one would "normally" expect to find in space. This will make it easier to describe deviations from the norm. Every point on a *normal surface* shall be the

Chapter 1   Descriptive Topology   7

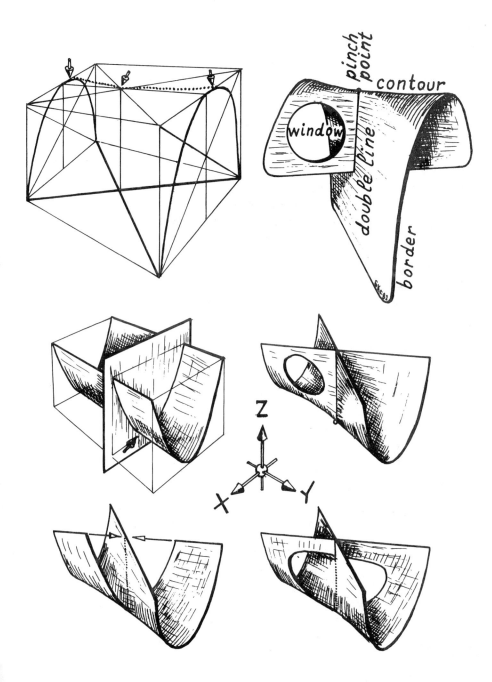

Figure 3   Whitney Umbrella

center of a spherical neighborhood wherein the surface is the image of a mapping *(X,Y,Z) = F(r,t)* from the plane to space. The *X,Y,Z* are (local) coordinates in the *target space*, *r,t* are coordinates in the *source plane,* and *F* is a *smooth, proper* and *stable* map from one, or at most finitely many, disjoint discs in the source plane. Recall that the components of a *smooth map* have arbitrarily many continuous partial derivatives. Every limit point *P'* in the target of a sequence $\{F(P_i)\}$ of image points of a *proper map* is itself the image, *P' = F(P)*, of a limit point of the sequence $\{P_i\}$ in the source.

The precise definition of a stable map, see Golubitsky-Guillemin [1973, p42 and 72], depends on the notion of the *Whitney topology* on the set of all smooth mappings between the given source and target. Informally speaking, let $\delta(P)$ be a continuous assignment of positive real numbers to all the points in the source. Two smooth maps are δ-close if their values as well as the values of all of their partial derivatives at *P* differ by no more than $\delta(P)$. A *stable map* is one which differs from sufficiently nearby maps only by a diffeomorphism in the source and in the target.

For present purposes it suffices to know what the image of a stable map of a surface into space looks like in the neighborhood of each point. If no neighborhood, no matter how small, of a given point looks like a mildly bent disc, then it is a *singular point*. A stable map can have three kinds of singular points. In a neighborhood of a *double point* a surface looks like two sheets of some fabric crossing along a so-called *double curve*. A neighborhood of a *triple point* looks like three surface sheets crossing transversely. Thus triple points are isolated. You can see why a quadruple point is *unstable*. A slight perturbation of one of the sheets would make four sheets cross each other so as to produce a little tetrahedral cell. Double curves are either closed, extend to infinity, terminate on the border or simply end at very special points, called *pinch points*. The vicinity of a pinch point looks like a *Whitney umbrella*, 3(12). The arrows in the line pattern, 3(11) point to important tangencies of the contour of a Whitney umbrella. This is the third kind of stable singular point. The stable maps comprise a dense and open subset (in the Whitney topology) of the space of smooth maps of surfaces into space. This is why such maps are considered to be "generic".

A normal surface admits a continously turning tangent plane in the following sense. For a point *P'* other than a pinch point, and each of (up to three parameter pairs *(r,t)*, the tangent vectors $\partial F/\partial r$ and $\partial F/\partial t$ at *P' = F(r,t)* are linearly independent and hence span a plane. At triple points there are three, along double curves two, independent normal directions to the surface.

Matters are more complicated near a pinch point, but can readily be described in terms of the umbrella's canonical form, $(X,Y,Z) = (rt, r, t^2)$, which looks like 3(22). As *r* moves through zero the parabola $r^2Z - X^2 = 0$ in the vertical plane *Y = r* squeezes through the double line and opens up again as its plane moves forward along the *Y*-axis. Or, in time *t*, the line *X − tY = 0* in the horizontal plane $Z = t^2$ turns as the plane moves down the Z-axis and back up again. The umbrella is also a ruled surface.

It is now clear that there are many limiting positions for a plane tangent to the umbrella at points approaching the pinch point. Detail 3(31) shows how to make a paper model by bending a square with a slit and "closing" the slit across the sheet. 3(32) has a window which "removes" the pinch point. Note that the pinch point is on the border of this window and the window cannot be "continued" past the pinch point because the surface bends in opposite directions there.

In the language of topology, all or part of a surface is *embedded* if it has no double points. All of 3(31) is embedded, as is the plane or the parabolic cylinder separately in 3(21). It is *immersed* if every point on it is in the interior of at least one disc which is embedded in the surface. This is the case at all points of the umbrella except the pinch point. 3(21) and 3(32) are immersed surfaces. We shall investigate these ideas in a more leisurely way in the fifth chapter.

CAYLEY CUSP. *Figure 4.*

There are two topological reasons for adopting normal surfaces as the basic forms for drawings. A sufficiently small distortion of a stable mapping of a surface can be returned to its original shape by an *isotopy* of the ambient space. In other words, there is a one parameter family of coordinate changes which removes the distortion. Moreover, arbitrarily near any smooth mapping there is a stable approximation to it. A practical way to check that a certain surface feature is unstable is to remove it from the surface by means of a small perturbartions of its parametrization.

Singularity theory also applies to drawings of normal surfaces in the picture plane. A mapping F of a surface into space, followed by a projection to the picture plane, becomes a surface-to-surface mapping, $(x,y) = \Phi(r,t)$. There is a technical difference in the notion of stability if you regard Φ as mapping the rt-parameter plane directly to the xy-picture plane, or as "factored through" a projection of a surface in space. But for practical purposes we can ignore this distinction.

Wherever the surface in space bends away from view, my line of sight moves along a curve of tangency, called a *contour*. The map Φ has a *fold singularity* on the contour. The contour can disappear from view in the following ways. It may pass behind a nearer part of the surface. When this happens, the contour should be drawn transverse (not tangent) to the edge of the nearer part, be that a border or itself a contour. It may reach a border or a double curve of the surface, in which case the contour disappears tangentially on that curve. The little arrow in 3(21) points to an example. These are the most difficult junctions of two different kinds of curves to

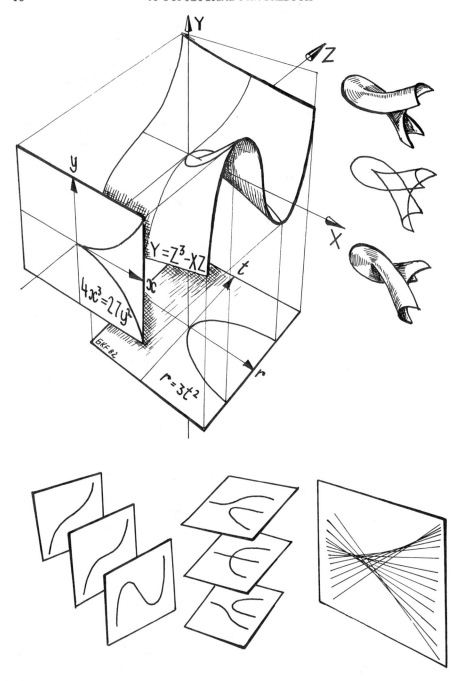

Figure 4   Cayley Cusp

draw in a picture. To help the eye, imagine that the surface has finite thickness. This way, contours approaching the picture plane merge into borders at a change of weight in the curves. There is a pointilliste style of drawing, favored by Tim Poston and Ian Stewart [1978] for example, in which a surface is represented by a uniform distribution of points on it. The contours are formed by an abrupt change of point density in the picture. Poston's style was very useful for my drawing 2:12(12).

Finally, the contour can itself turn away from view at a *cusp*, so called for the shape of its projection to the picture plane. The canonical form of this singularity is illustrated by 4(11). The cubic surface, $Y = Z^3 - XZ$, admits the parametrization $F(r,t) = (r, t^3-r\,t, t)$ if you identify X with r and Z with t. Orthographic projection to the xy-picture plane produces the surface-to-surface map $\Phi(r,t) = (r, t^3 - rt)$ whose determinant of first partials (the Jacobian) is:

$$\frac{\partial \Phi}{\partial (r,t)} = \begin{vmatrix} -1 & 0 \\ -t & 3t^2 - r \end{vmatrix}$$

This mapping fails to be regular along the parabola, $r = 3t^2$, where the Jacobian of Φ vanishes. The image of this curve of critical points in the picture plane, $(x,y) = (3t^2, -2t^3)$, is the semi-cubical parabola, $4x^3 - 27y^2 = 0$. This curve was first studied by William Neil in 1660, who found its length by infinitesimal methods at the dawn of the calculus. That the location of a contour on a surface depends on your viewpoint is evident in this "picture of a picture," 4(11).

Consecutive sections by parallel planes of the cubic surface, in all three directions, have interesting 2-dimensional interpretations. Planes orthogonal to the X-axis 4(21) show the elementary catastrophe of a function with two extrema changing to a function with none. In the horizontal sections orthogonal to the Y-axis, 4(22), you see the associated *pitchfork bifurcation* of mechanics. The vertical sections parallel to the picture plane are straight lines, 4(21), whose envelope is a cusp. It is also a ruled surface.

The last example is important for reconstructing the Cayley surface from its line pattern in the picture plane. The detail 4(12) shows a right-handed (top) and left-handed cusp (bottom) based on the same line drawing (center). The cusp has a field of nonzero vectors, $\langle 1, -t \rangle$, tangent to it. Pull this structure uniformly in the time direction and stretch the vector by the scalar factor $s = r - 3t^2$ to obtain this mapping into 3-space: $F(s,t) = (3t^2, -2t^3, t) + s\langle 1, -t, 0 \rangle$. This differs from the earlier parametrization of the cubic surface by a change of coordinates in the parameter plane, $(s,t) = (r-3t^2, t)$. It suggests a way of bending the Cayley surface (by an isotopy) into the shape of a standard saddle. Deform the X-coordinate by replacing it with $X - hZ^2$. As h goes from 0 to 1, the expression $Y + XZ - Z^3$ goes to $Y + XZ$.

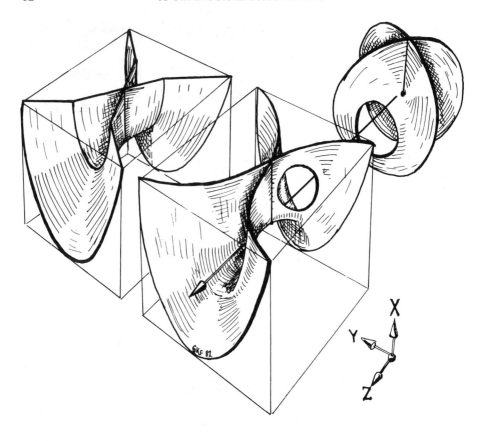

PINCH POINT/BRANCH POINT   Figure 5

These then are the elements of descriptive topology: normal surfaces with their border curves and double curves, triple points and pinch points, and their pictures with cusps and contours. Most of the visually stable relations between pairs of these elements have been mentioned. You should investigate the rest. For instance, how should a curve that crosses a contour on the surface be drawn on the picture plane? Here is another one that is often missed.

In general position, pinch points are located at ordinary points on the contour. To see this, parametrize a neighborhood of a contour point by $F(r,t) = (r, t^2, u(r,t))$ and recall that the normal vector on a Whitney umbrella vanishes only at the pinch point. Compute the normal vector by taking the cross product of the two tangent vectors to the surface:

$$\frac{\partial F}{\partial r} \times \frac{\partial F}{\partial t} = \langle -2tu_r, -u_t, 2t \rangle.$$

Thus, if there is a pinch point in the range of F then it occurs when $t = 0$, hence it also lies on the contour.

This rule is almost always broken in conventional representations of branch points as they occur in graphs of complex functions. How branch points and pinch points are related, in particular that the shadows of Riemann surfaces such as $z = \text{Re}\sqrt{(x+iy)}$ have them, can be seen as follows. The isotopy $(X(t),Y(t),Z(t)) = (X + tZ^2, Y - \frac{1}{2}tYZ)$ takes the graph $(X,Y,Z) = (u^2-v^2, 2uv, v)$ to the canonical umbrella, (u^2, uv, v) at $t = 1$. The composite picture shows the conventional view of a branch point to the rear, 5(13). This is not a visually stable picture. There should be a small, 3-cusped contour in the shape of a *deltoid*, characteristic of the *elliptic umbilic catastrophe* of René Thom [1972]. If you continue the surface 5(12) upward then its contour would close at the third cusp of the deltoid. The non-differentiable surface 5(11) shows a transition from the umbrella to the Riemann surface. It was the model for drawing the smooth surface in the center.

Here, and in the other pictures in this chapter, I have left some of the construction lines fully visible. How these may be laid out efficiently and converted to a picture of a surface in space is the subject of the next chapter.

2
METHODS AND MEDIA

This chapter is about a method I developed to draw the pictures in this book. It is, of course, an amalgam of many techniques which I learned from other illustrators. My earliest efforts at drawing surfaces were inspired by catastrophe theory. Many of the models proposed by Thom and Zeeman were purely descriptive: a figure of speech based on the geometry of a polynomial. The technical economy and operational simplicity of such models are very appealing. Accordingly, I shall present my drawing method in terms of such a geometrical metaphor. If you are familiar with the story of the calculus, as told by Carl Boyer [1949] for example, you will also recognize my debt to the schoolmen Suiseth and Oresme. In their theory of forms, a precursor of Cartesian geometry, they graphed an inexact intensity on the latitude (=ordinate) against the longitude (=abscissa) measuring some duration or physical extension.

CUBIC METAPHOR. *Figure 1.*

The form of the model is that of the cubic function $\ell = t^3 - 3rt$ on the interval $t_{min} \leq t \leq t_{max}$. The coefficient 3 and the origin $\ell = r = t = 0$ have no particular significance except to simplify the arithmetic. The latitude $\ell - \ell_{min}$ qualitatively measures the density of lines drawn on the picture plane at any moment, starting with the *reference frame* at $\ell_{min} < 0$, and ending with the finished picture at $\ell_{max} > 0$. The reference frame is some familiar figure (cube, cylinder, sphere, torus) or, for very simple pictures, just the outline of the object. As yet, there is no good definition for the *line density* measured by the latitude. It records the intensity of subjective impressions, such as regarding two lines of unequal thickness as being of less weight than two lines of equal thickness.

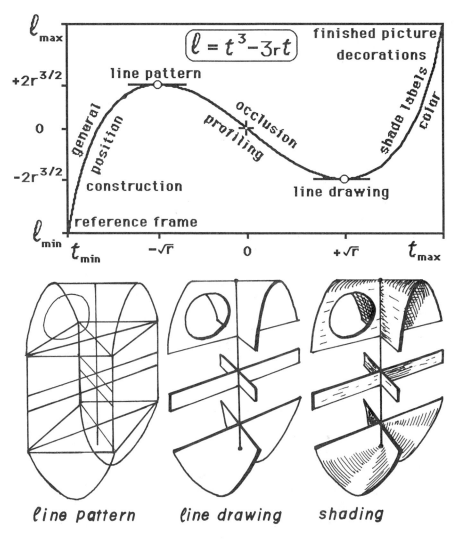

Figure 1 CUBIC METAPHOR

Successive stages in the production of a picture are measured by the duration $t - t_{min}$. The control variable r relates to the overall effort needed to complete the picture. Easy pictures are coded by the uncomplicated cubic for which $r \leq 0$. The inflection point at the origin represents a picture which is just adequate for its purpose. Points to the right, hence further up the graph, represent pictures with additional *decorations*, such as shading, cutting windows, labelling, coloring, etc. Decreasing the control parameter r shortens the total time $t_{max} - t_{min}$ it takes to complete the drawing.

The interesting pictures have the interesting cubic $r>0$ as their model. On the Cayley surface, 1:4(11), we have moved forward to the right. From the derivative, $\frac{d\ell}{3dt} = t^2 - r$, you see that the number of lines added to the frame increases at a positive rate until the *line pattern* is reached at the relative maximum. In addition to construction lines, this pattern includes all borders, contours, double lines, triple points, cusps and pinch points, windows and lines that eventually guide the shading. All these lines and points should be in general position, if at all possible, to avoid visually ambiguous coincidences later on. The pattern also serves as an underlay for tracing the picture. When a sequence of pictures illustrates a changing object, the pattern serves as a template. The visual continuity achieved by using a common line pattern assures that successive pictures differ only in the most relevant aspects.

In the next stage the line density decreases again. The construction lines are removed from the pattern. Hidden lines are either omitted, drawn dotted or dashed, or the visible lines are enhanced by thickening them relative to the hidden lines. This is sometimes called *profiling*. You may wish to stop at the inflection point of the cubic model, leaving some extra lines as guides to the eye. Otherwise, proceed to the relative minimum, which represents the *line drawing*. A good line drawing has the minimum number of lines still adequate for recognition. It is, in a sense, the best for mathematical illustration because it leaves the most space for decorations to be applied by the user.

As the design of a particular picture becomes simpler, more stylized and more memorable, the preparatory steps tend to be omitted and a line drawing is produced with a few strokes of the pen. Watch a knot-theorist draw a trefoil knot with three quick unerring lines. It is tempting to apply Zeeman's dialectic for catastrophe theory also to my cubic model for drawing. Consider, for instance, the hysteresis loop. Often, already in the earliest, exploratory pencil sketches there is a jump from a messy line pattern to a first approximation of the picture, skipping a formal line drawing. As I add line upon line to a frame, with crude shading to simulate curvature and to suppress hidden lines, I will suddenly see a satisfactory design. I ink this in before erasing the construction lines, lest the picture vanish also. I then trace a first approximation for a line drawing. If it proves unsatisfactory, I jump back to designing a better line pattern and the loop is complete.

BASIC SURFACE PATTERNS. *Figure 2.*

A happier (anastrophic!) discontinuity occurs on reversing the hysteresis. A mere outline sometimes suggests an immediate jump to a good line drawing. An example of this is John Stillwell's convention for drawing cusps, for example see page 57 of his text [1980]. An oval outline 2(11) becomes an

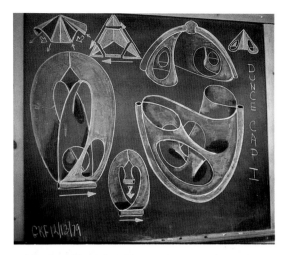
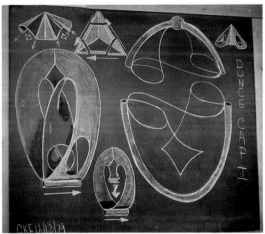

PLATE I. *Blackboard Dunce Cap.*
Available light photograph on high speed Ektachrome daylight slide film.

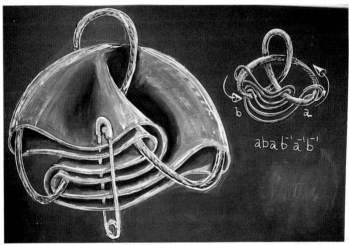

Plate II. *Diapered Trefoil Knot.*

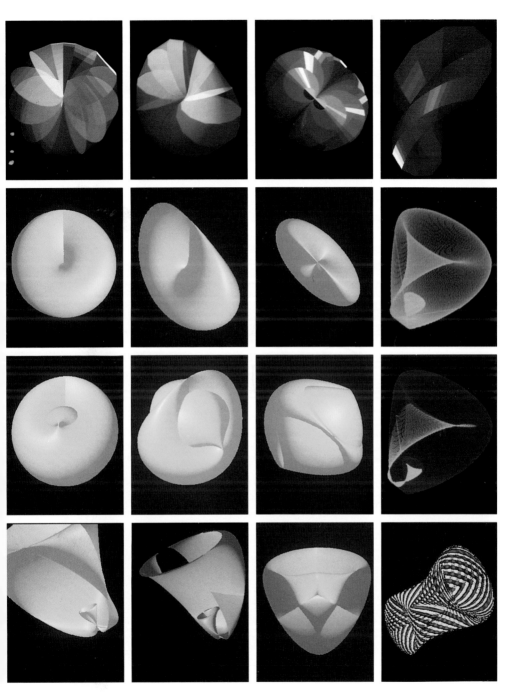

PLATE III. *Cross Cap and Limaçon.*
Donna Cox and Ray Idaszak, Electronic Imaging Laboratory and National Center for Supercomputing Applications, University of Illinois.

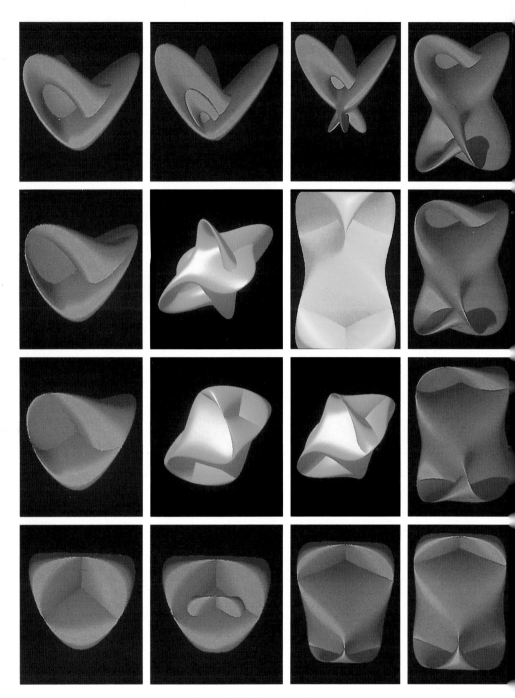

PLATE IV. *Etruscan Venus.*
Donna Cox, George Francis and Ray Idaszak, National Center for Supercomputing Applications, University of Illinois.

Figure 2   BASIC SURFACE PATTERNS

elegant torus 2(21) with just two additional lines. In a line pattern, 2(12), the contour for the hole in the torus is a closed curve with two swallowtails. Hiding the invisible cusp of each swallowtail produces a different line pattern, 2(13). Note how the longer of Stillwell's contours in 2(21) continues beyond the cusp points, hinting at a shading scheme, 2(23). The convention works well for characteristic views of the Klein bottle 2(22), Boy surface 2(31), Morin surface 2(32) and Roman surface 2(33). You should "mold" these line drawings by shading with a pencil, "pushing back" the regions near the cusps as outlined by Stillwell's curls.

TRIPRONG. *Figure 3.*

In designing a line pattern for a surface one is sometimes able to place the border curves with confidence, only to wonder where the contours should go. To obtain a plausible answer I smooth a piecewise flat and cornered version of the surface. Each line on such a *polyhedral surface* is an edge bordering one face, or separating two or more faces. Because it is flat, each

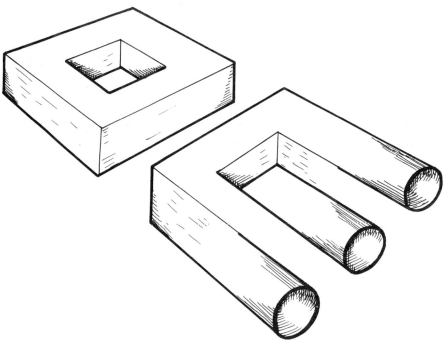

Figure 3   TRIPRONG

face of the polyhedron is either wholly visible or wholly invisible. A simple edge separates only two faces, one of which might be hidden. To distinguish an edge whose other face is hidden from an edge which merely separates two visible faces, drafting teachers recommend heavier lines for the former relative to the latter. Let me call them *contour edges* and *face edges* of the polyhedron. The graphical convention simulates a phenomenon of practical optics. As a real object is brought closer to the observer for a better view, or his eyes scan a stationary object, contour edges produce stronger retinal stimuli than face edges. This is so because they separate foreground from background, which have a stronger light contrast than the difference in albedo for contiguous visible faces. As the corners of a piecewise flat surface are smoothed, contour edges become contour lines, but face edges disappear from the line drawing. They are useful as shading guides, however.

An amusing optical illusion, the *triprong*, neatly summarizes this lesson. A profile analysis exhibits three pairs of contours tangent to the border circles. Only four of them are contour edges of the polyhedron. Two of these switch visible faces, which is impossible. Two more contour edges illogically become face edges.

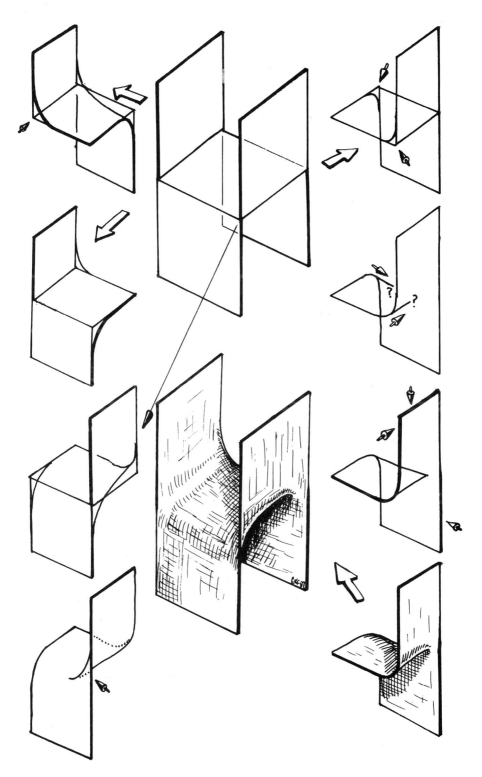

Figure 4 Rounding a Polyhedral Saddle

Rounding a Polyhedral Saddle. *Figure 4.*

An example involving nontrivial contours is this graphical transformation of a piecewise flat saddle, top center, to a smooth surface spanning the same border frame, bottom center. The square, horizontal face of 4(22) has two face edges (left) and two contour edges (right). The right column shows how the confluence of these two contour edges becomes a cusp. First, isolate a detail of this corner, 4(13), and bend the (new) borders. This initiates two contours, 4(23), which, in this case, cannot merge smoothly into a border or into each other. Hence they meet at a cusp, 4(33). For simplicity and optical coherence of the various parts I have not used perspective in this drawing. However, line patterns based on boxes in affine projection suffer from the Necker cube illusion: "Which is front and which is back?" Thickening the facing borders, as in 4(33) helps decide which view is intended, 4(43).

The other three corners, left column, present fewer graphical problems. No contours are generated when two face edges meet, 4(21). Only one contour is generated in 4(11) and 4(31), and it can merge with a border in both cases. In provisional sketches I sometimes deliberately replace such contour/border junctions by a cusp, 4(41), to avoid Escher-like impossibilities.

Dunce Hat. *Figure 5.*

The technique of piecemeal graphical construction is also useful when you don't already know what the final object will look like. Designing a *dunce hat* is a good example. This topologically polyhedral surface is the abstract cell complex with one vertex, one edge and one face, obtained by identifying the incoherently oriented sides of a triangle, 5(11). Two of the three pairs of edges in this triangle match easily to form the conical hat, 5(12), commonly associated with the great schoolman, Duns Scotus. The topologist, however, also expects you to glue the circular rim to the straight seam.

What the dunce hat might look like is not at all obvious from this prescription, nor that it lives in 3-space at all. Our first visualization uses only common topological ingredients. Suppose you tilt the seam, 5(13), into the same plane as the rim, thereby drawing out a triangular flap or *sail*. Wrap this around the cone, 5(14), and finish the gluing, 5(21). The resulting 2-complex, 5(22), is not the dunce hat. It has two extra edges: where the sail meets the cone, and the free edge of the sail. For a dunce hat, imagine the sail to be of double thickness, open to the cone but sewn on the free edge. By blowing into the base of the cone, you can inflate the dunce hat to look something like 5(32). You arrive at a similar shape by bending the cone, 5(12), so that the vertex becomes the cross in a planar figure eight,

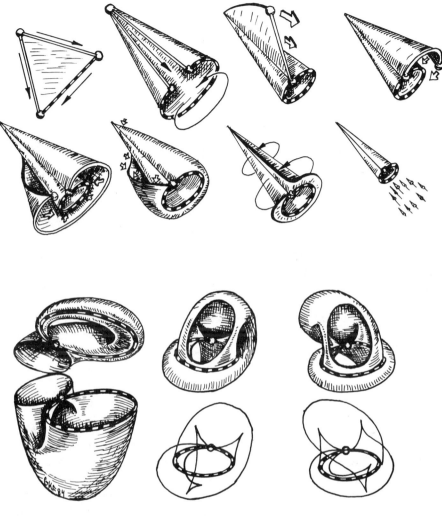

Figure 5   DUNCE HAT

5(31). (Bend it the other way to match 5(32).) A standard graphical construct, the hemisphere pinched into a hemitorus with zero inside diameter, 5(41), connects the two loops of the border.

Now it is clear how the dunce hat is a *spine* for the 3-ball. You could add layer upon layer of paint, spreading it more thickly inside the snail-like hollow so as to fill it up more quickly, until the surface is built up into a smooth, round ball. Technically speaking, the ball can be *collapsed* to the dunce hat. The dunce hat itself, however, is clearly not collapsible, there is no place to start chewing away on the edge of a face. The false dunce hat

5(22) with the one ply sail, is collapsible. Imagine it to be made of improperly flame-proofed paper. Start it smouldering along the free edge of the sail. It collapses first to the cone, 5(23). This, in turn, collapses to its vertex, starting from the free edge along the rim, 5(24).

Though the entire biography of the dunce hat does not fit into this chapter, a few of its properties are worth mentioning. It is a contractible but not collapsible spine of the 3-ball. It is, however, 1-*collapsible,* which means that its Cartesian product with a "1-ball" is collapsible. You should think of this product as a thickening of the surface into the fourth dimension. Zeeman [1964] conjectured that every contractible 2-complex is 1-collapsible. Marshall Cohen [1977] showed that higher dimensional analogs of Zeeman's conjecture are false. Gillman and Rolfsen [1983] proved the logical equivalence of the conjecture for "standard" spines to the celebrated Poincaré conjecture. The curious structure of the dunce hat near its vertex makes it an example of a "non-standard" spine of the 3-ball.

I am indebted to Marshall Cohen for suggesting the construction 5(31) plus 5(41), and how the dunce hat D contracts to a point. Place D inside a ball B^3 centered at the vertex V. Let $R:B^3 \to D$ denote the retraction that collapses the ball to its spine. Let $G_t(X) = (1 - t)X + tV$ denote the radial contraction of the ball. The composition of G followed by R contracts D to V. To imagine the route on D a point X of D takes, just retract the radial line XV onto D and contract the two curves in synchrony.

Duns Egg. *Figure 6.*

A contour line drawing, 5(42) or 5(43), for the snail shaped dunce hat is too complicated to remember. Here is a more systematic way of designing a dunce hat with better symmetry as well as a simpler contour, 6(31). Moreover, it illustrates, in dimension two, a particular way of cutting and pasting topological objects which is even more useful in three dimensions. Isolate a neighborhood of the vertex by detaching three triangles, 6(13). Assemble these separately into the cone over the *vertex link,* 6(12) of the complex. The edge pair opposite cuts 2 and 3 on the residual hexagon glues together like that of a rectangle for a Möbius band, 6(23). The edge opposite cut 1 now draws down like shade, but twisted to align the edges, 6(22).

There remains the visualization of how the vertex link, 6(12), could possibly fit 6(22). I arrived at the egg shaped solution, 6(11) plus 6(21), something like this. Push the small loops (cuts 2 and 3) out from the plane of cut 1 by changing the contour cusps from 6(42) to 6(41). Next, bend the edge sharply into the keel of a hemisphere to bring the border into the equatorial plane, 6(43).

CHAPTER 2 METHODS AND MEDIA

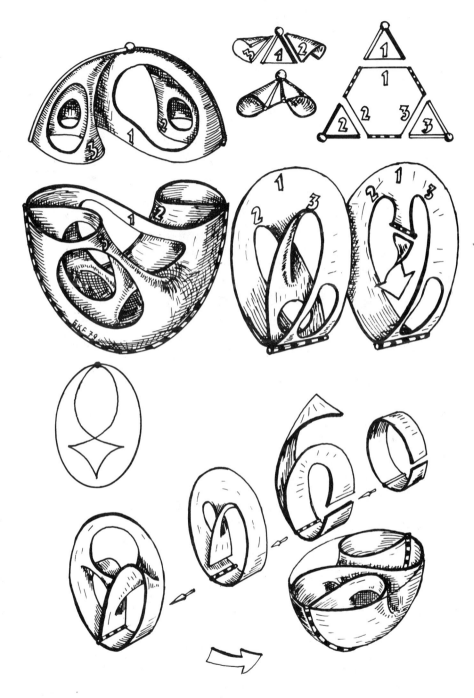

Figure 6 DUNS EGG

MÖBIUS BAND. *Figure 7.*

This picture shows two solutions to the the problem of visualizing a Möbius band whose border is a plane circle. It also illustrates two different graphical techniques for constructing such visualizations. Both begin with the Möbius band regarded as a disc with a twisted handle, 7(12), whose border looks like a (nearly) plane figure eight. A deformation of 7(12) to the usual shape of the Möbius band, 7(32) is indicated by the transitional form 7(22).

The 3-dimensional approach for modelling a given shape is on the left. Use standard surface pieces which are easy to draw and easy to describe verbally. Here, 7(11), a hemisphere is attached to a disc with two quarters bent in opposite directions away from a radial slit. The border of this object spatially resembles the twisted figure eight of 7(21). You can take out the twist by turning the semicircle through two eighth spheres, 7(21). The border, now a semicircle and its diameter, shrinks to a plane circle, 7(31). The closed surface you obtain by sewing in a disc is a Steiner cross cap, about which I shall have more to say in Chapter 5.

Although 7(21) is easy to make out of a bowl and paper, it is difficult to draw convincingly without some shading. I have, however, left the interior of the shell unshaded to heighten the contrast. The wire, that wends its way through the surface as seen through a window, is also meant to enhance the sense of spatial extension.

A 2-dimensional approach is shown on the right. First reduce the number of layers by sending a point on the surface to infinity. Think of a stereographic projection from the south pole of 7(12) to the paper plane. More directly, consider a plane minus a disc and connect opposite sides of this disc with a half-twisted strip, 7(31). To bring the figure eight shaped rim into a plane, you need to untwist one of the two loops, pulling the surface along with it. The result of this deformation is equivalent to sewing in a Cayley cusp, as shown by the arrow. With two windows, the result looks like 7(33). Its anatomy is depicted by 7(32). The two pieces of this figure do not fit together directly. You must first pull the relevant border arcs into a plane perpendicular to the picture plane. The surface 7(33) resembles Dan Asimov's "Sudanese Möbius Band." This is a computer graphics film of a Möbius band expressed as a minimal surface in 4-space.

INK AND PAPER.

This is a good place to say a few words about the drawing equipment which I have found useful. A soft pencil, a pad of blue-lined bond and a kneadable art gum eraser head the list. The blue lines are a handy guide for the horizontal direction and won't reproduce on most copying machines. The

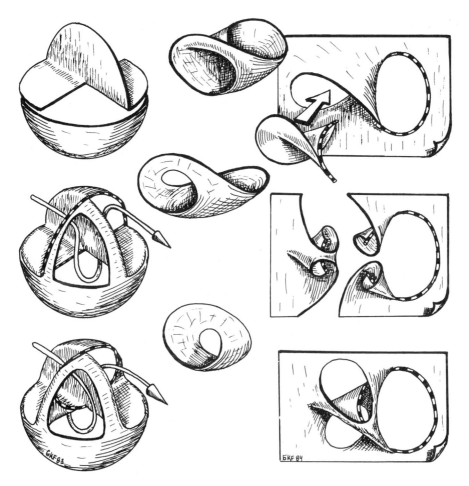

Figure 7　　　　　　　　Möbius Band

art gum has an unlimited capacity for graphite, won't leave little bits of eraser around, and doubles as instant modelling clay. Next come inkpen, plastic clipboard and lots of tracing paper. The ink should be black, the clipboard transparent, and the paper can be typist's onionskin.

For a publishable copy, india ink on high rag content tracing paper (vellum) works very well. A set of drafting pens is worth the investment despite the nuisance of keeping them clean. For minor corrections I apply white watercolor with a very fine brush. A small drawing board, T-square, clear plastic triangles, and an ellipse template complete the basic list. Use drafting tape instead of thumbtacks to hold the paper to the board. For small circles, a template is more convenient than an inking compass. Though tricky to use, splines (french curves) are a blessing for unsteady hands. To insure continuous curvature, it is a good idea to fit a french curve to three points, two of which match those of adjacent triplets. Beyond this, the sky is the limit and art supply stores are a good source of liberal advice on advanced tools such as lettering guides, pencil sharpeners, electric erasers and the like.

Reduction of up to 50 percent produces flattering prints. For this reason I generally use a size 0 or 1 pen for ordinary lines, 2 or 3 for borders and lettering, 00 or 000 for shading and construction lines, all with fairly good results.

Chalk and Blackboard.

Yuri Rainich, who first introduced me to the wonders of projective geometry, held that the presence of rotations in the Euclidean group of congruences reflects the Greek habit of drawing figures on the ground as the students stood about in a circle. Be that as it may, the vertical blackboard is now the mathematician's prime medium of pedagogical expression. No amount of explanation accompanying a complete figure on a page can match the information transmitted while creating the same figure at the board, talking all the while. The ease of adding and subtracting detail, of correcting errors and amending diagrams are familiar to all math teachers. Trying to take notes of such blackboard artistry can be a trying experience for the audience. All too often the teacher's haste or deficient technique prevents the student from recognizing the object altogether. To help students record enough of a figure to redraw and complete it at home I generally construct the blackboard drawing from scratch rather than reproducing the finished work from memory or my notes. I draw the reference frame to establish a stereometric context for the viewer. It declares my intention of sketching something in space. The line pattern, committed to memory, not only helps me avoid common errors of perspective and tangency, but also presents the recipe for reproducing the figure. Its shape is further clarified by erasing the hidden lines. A window or two restores a deleted detail. A highlight drawn with the flat of the chalk strengthens the appearance of curvature. A fair command of these techniques should be as common a virtue among topologists as their glib command of the language of sets and the notation of symbolic logic.

The kind of picture to draw on the board depends, of course, on its purpose. In the first instance, the lecturer wants a *pictograph* to accent the exposition. Here speed and simplicity are essential, while artistry is not. The exposition could proceed quite well without the picture. Simplicity in a pictograph need not mean poverty of imagination. The particular object selected should evoke a rich set of associations in the audience. Compare, for example, two possible figures to go with the phrase "Let E be a fiber bundle over base B and fiber F ...." The customary rectangle E above a segment B with a vertical line F above a point on B illustrates the local product structure but misses the chance to point to global topology. A Möbius band takes only one extra line to draw. Labeled E, with its equatorial circle as B and a crosscut for F, this becomes a good pictograph for a non-trivial fiber bundle.

In the second instance, a *working figure* is part of the exposition, and substitutes for a cumbersome, set-theoretic description of a topological procedure. Current taste in mathematical rigor prefers text which dispenses with pictures. A good working figure is useful for composing such rhetoric and keeping track of its logic. Because detail is added piecemeal to a working figure, graphical accuracy and visual clarity are essential here.

In the third instance, a complete, *polychrome picture* is drawn on a blackboard with colored chalk. Though time-consuming and difficult, this medium has its own charm and utility. An elaborate picture drawn on a secondary blackboard is a good advertisement for a seminar talk. Later, as my working figure on the main board deteriorates in the heat of the lecture I refresh my audience's attention by referring to the picture off to the side. Once a design is completely worked out in private, a full-color version can be reproduced fairly quickly on demand. For such a picture I go through the usual steps of framing and placing the line pattern into general position. The sheets of the line drawing are colored in solidly with the flat side of the chalk. Lecturer's chalk, though dustier than regular colored chalk, produces richer tones. Don't leave such drawings on a board which is not real slate. The dyes in the chalk tend to make such drawings semi-permanent. I then go over the lines once more. Black chalk is very effective for shading and for establishing a strongly contrasting background. White chalk is for the highlights. It is surprising how little color needs to remain for a field to be perceived as uniformly painted. Photography is the only way to preserve such fragile works. To make sure that you have some record use a Polaroid camera. But 35mm slides are more satisfactory and can also be shown to a larger audience. Among the color plates in this book are some examples. Three photos, of an early version of the Duns Egg 6(11), show the line-drawing stage, the monochrome version, and a polychrome picture. I used color as a way of labeling matching surface features.

For preparing and presenting polychrome pictures, 35mm slides have certain obvious advantages over transparencies. For their preparation I only need a slate blackboard, high-speed film and reasonably quick photo developing services. During the talk I can look at and point to the same image as the audience sees. I don't have to switch between looking at the brightly lighted top of the transparency projector and the darkened room. The blackboard is an invaluable medium also in the design stage of a picture. The large scale of chalk drawings and ease of correcting them greatly helped me solve the problems in descriptive topology whose solutions you see in this book. Sometimes, the time it takes to draw the picture during the lecture is also useful. The three photos of the *Diapered Trefoil Knot* have a little story worth telling in this regard. It began when Bill Thurston asked me to draw the disc spanning a given curve in the complement of the knot. The curve linked the knot so as to represent the relator in a Wirtinger presentation of the knot group.

Though I cannot review knot theory here let me remind you how to read the relator. You can find out more about knots in the classic text by Crowell and Fox [1963]. Orient the knot projection and label the segments between consecutive under-passes. (The trefoil has three of these, but one label is superfluous.) Choose a base point in the complement of the knot and draw an oriented loop from the base point around each of the segments except the superfluous one. Curl your right hand around the knot segment with fingers pointing along the loop. The loop label receives a positive (negative) exponent if your thumb points in the same (opposite) direction as the knot. In this way any word in the free group generated by the segment labels determines a closed curve in the knot complement. Since knot groups can be presented with a single relator there will be one word whose corresponding curve is spanned by a disc embedded in the complement of the knot.

The solution to the problem is shown on the right of the first photo. On the left is an intermediate stage of the disc which I drew to help visualize the former. Serendipity, however, intervened. The helper picture already solves the problem if you use a conjugate of the given relator. This is done in the second picture. The safety-pin serves as the base point. As long as the diaper is pinned, it cannot slip off the knot. This picture was now simple enough to remember and reproduce on demand. The green one in the third photo was drawn in "real time" in Benno Artmann's seminar in Darmstadt. With all the practice, it took less time to draw it than to explain the topology, the group theory and the descriptive topology involved.

SLIDES AND TRANSPARENCY.

For most public mathematical presentations, especially with a time limit and a need for perfect illustrations, there is nothing to replace the trusty overhead projector. This medium tolerates a great deal of expository ingenuity and individual freedom. In preparing a transparency, you have some choices to make.

Despite technical advances, xerographic copies of your ink or pencil drawings on the plastic sheets usually require some touching up. Drawing ink works fine, and the picture can be enhanced further with the judicious use of colored inks. These come in two kinds, the water soluble, and thus eraseable, and the permanent ink markers. A combination of the two is useful if the transparency is to be reused. Ordinary india ink can also be erased by just rubbing. Use razor blades to cut out, not scratch out, hopeless mistakes. An overlay will patch the hole nicely. I find it convenient to put the most complicated parts of a figure on the transparency before the lecture, and add some details to it, or on an overlay, during the talk. Remember, you can work on both side of the plastic.

A figure can be drawn directly on the acetate, using a line pattern or complete picture as an underlay. Ordinary drawing ink works; inks more specifically intended for drawing on film are more permanent. They contain an acid but can be erased with a suitable fluid, at least for a while. This is usually neater than xerography, but more time-consuming. Since the solvent of many inks will smear adjacent pigments already applied, I sometimes color on alternating sides of the sheet. When using several overlays, it pays to tape the sheets together, so that they fit correctly.

COMPUTER AND DRAFTING TABLE.

Computer-generated visual displays certainly are marvelous aids to the imagination. This spectacular new medium has been used by Tom Banchoff [1977] and Nelson Max [1977] to produce films of great beauty and lucidity in geometry and topology, respectively. Eventually, the graphics computer will be as important a tool in descriptive geometry and topology as india ink and slate blackboards. But not yet. Nelson Max's hidden-line, fully tinted and shaded animation of Morin's sphere eversion is a masterpiece of computer graphics. It also required the best available hardware and programming skills of its day to produce. Like the moon landing now two decades past, Max's expensive effort has had no sequel. It will be quite some time before such technology is available to the journeyman topologist.

Tom Banchoff's *wire frame* images of surfaces are technically less demanding. Easier to program and simpler to display, they are also less easily visualized. Here all of the generating curves are fully visible. The illusion of depth is restored somewhat by letting the surface rotate steadily as it undergoes deformations. Even so, the passivity inherent in watching the fixed schedule of a motion picture, videotape or computer demonstration hinders the understanding of most viewers. The task of subsequently xplaining to students what exactly they did see in the films by Banchoff and Max led me to invent the topological picture stories in Chapter 5 and Chapter 6 respectively.

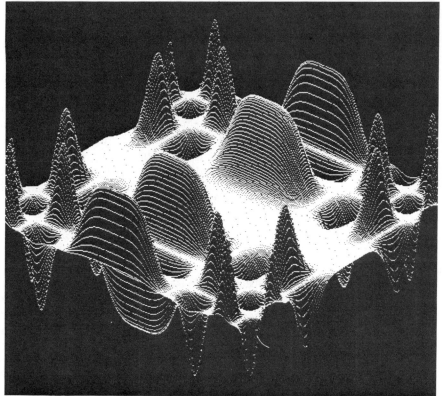
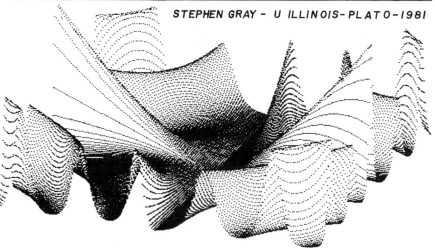

Figure 8   FUNCTION GRAPHS

Function Graphs. *Figure 8.*

The *hidden line problem*, otherwise easily solved by eraser or tracing paper, becomes a nontrivial obstacle to either speedy or economical computer drawing. The algorithm used by Stephen Gray for these graphs of trigonometric polynomials is quite simple and works well though slowly in BASIC on an Apple. Originally, it took ten dedicated minutes in the dead of night on PLATO's Cyber73/CDC6500 timeshared mainframe. It was the convenience of the PLATO system and the uncommonly supportive personal environment that got me started in programming there rather than on a proper graphics setup. My computer novitiate thus was similar to what awaits current beginners in micro-computing. I developed a hybrid approach where the computer furnishes the underlay for a pen drawing. It still serves me well in the micro-computer world of reduced resolution and marginal printing quality.

Cancelling Pinch Points: Wire Frames. *Figure 9.*

The next three pictures are a computer-assisted exploration of how to cancel a pair of pinch points. The program uses Morin's parametrization of the homotopy, $(X,Y,Z) = (x, y^3 + (t - x^2)y, y^2)$. First come three wireframe stereographs corresponding to moving t from negative to positive through zero. You can isolate the parabolic double curve by factoring the equation $Y = 0$ into $y = 0$ and $y^2 = x^2 - t$. In the space coordinates these become $Z = 0$ and $Z = X^2 - t$. Thus the vertex of the parabola is at $(0,0,t)$.

A *stereograph* is a pair of images of the same object which looks 3-dimensional when each view is seen by the appropriate eye. Traditionally, the images are placed so that the left (right) eye looks at the left (right) picture through a prismatic instrument called a *stereoscope*. The stereographs here are meant to be viewed without a stereoscope and are therefore reversed: the left (right) view is on the right (left) of center. To see them you must cross your eyes and refocus. To learn this trick look at a pencil tip held about half way between your nose and the stereograph. Move the tip closer or further away until two of the four fuzzy images fuse. The center image will still be fuzzy at first. Don't try to focus on the page. Continue to look at the tip of the pencil, and think about the surface until the middle image becomes sharp. With a little practice, you should be able to visualize reversed stereographs on paper or the computer screen quite easily.

If you have a stereoscope or get a headache from crossing your eyes, just xerox, cut and paste the stereograph, reversing the right and left sides. As long as the center of the two images are no further apart than your eyes and the images do not overlap, you can visualize also the parallel stereograph

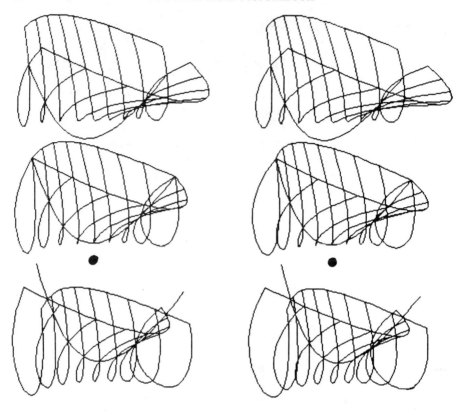

Figure 9 CANCELLING PINCH POINTS: WIRE FRAMES

unaided. Start by placing your nose right up to it and between the two centers. It is helpful to mark these as dots. Relax your eyes and let the (extremely) fuzzy images fuse. Now, slowly, draw back, keeping the them fused, until the middle picture comes into focus.

For unaided viewing, especially of a micro-computer screen, I prefer the crossed eye stereoptic method to the traditional one. Since the two images can be further apart, they can be larger. Crossed eyes also seem to be more tolerant of geometrical inaccuracies in the two images.

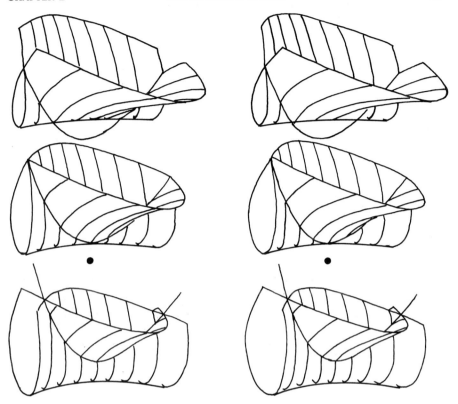

CANCELLING PINCH POINTS: HIDDEN LINES   **Figure 10**

Hand-tracing the visible parts of the curves in the previous figure and interpolating the contour lines produced these drawings. At the top the two pinch points are where the double curve, a parabola, leaves the surface. In the middle, the double curve just touches the exterior. At the bottom, the surface is immersed. It is curious that small inaccuracies in my tracing are more disturbing, when viewed stereoptically, than the wiggliness of the lines in the screen print of the wire frame.

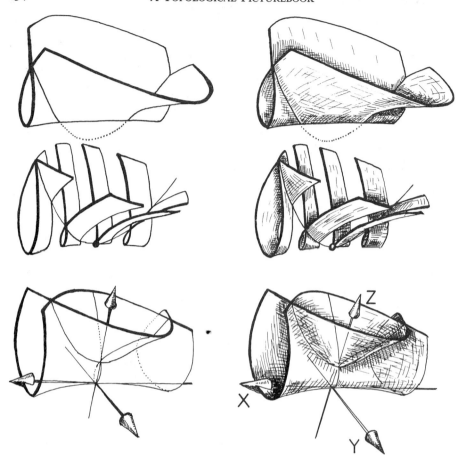

CANCELLING PINCH POINTS: ELABORATIONS   Figure 11

These pictures explore the various possibilities for achieving a sense of space without the danger of eyestrain. Detail 11(11) is a line drawing. The dotted part of the double curve is called the *whisker*. It is part of the algebraic variety whose equation, after eliminating the parameters, is $Y^2 - (Z + (t - X^2))^2 Z = 0$. Within the limitations set by the computer equipment I had at my disposal, the level of programming ability I had mastered and the inherent difficulty of the graphical problem to be solved, I developed this medium into a drafting-table tool, a template for shapes to be worked out by hand. Often the hand won.

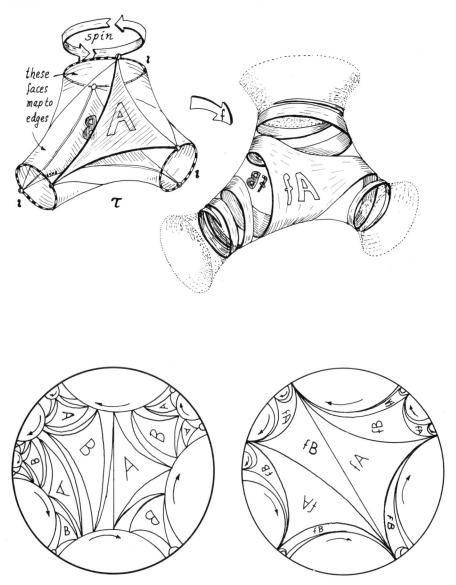

Spinning a Pair of Pants

Figure 12

Some components of my PLATO graphics editor served me well when it came to doing pictures for others. The hyperbolic plane routine automatically converts lines drawn in one model into any of the others. Those lines in the triangulation of a part of the Poincaré disc, 12(21) and 12(22), which went to the boundary were so converted from straight lines in the Beltrami-Klein model. It was quicker to complete the remaining arcs by hand. This illustration for a paper by Thurston [1986] solves an interesting graphical problem. 12(11) shows a triangulation of the thrice punctured sphere, or *pair of pants*, covered by 12(21). In the process of pushing everything on the border lines in the covering to one side, the borders of the Riemann surface are spun about themselves an infinite number of times. All but two of the triangles collapse to edges. The trick was to let the eventual position of the rear simplices, labeled *B*, occupy the space left over by the eventual position of the from simplex *A* . It was important to suggest that the surface continues beyond the belts but not so as to detract attention from the pants themselves. Tim Poston pointilliste style seemed most appropriate here.

LAMINATION AND EIGHT KNOT. *Figure 13.*

The hyperbolic plane routine was indispensible for designing this illustration for another Thurston paper [1982]. It is supposed to suggest filling up the two hemispheres by continually drawing more spherical triangles whose vertices are on the equator. Moreover, no two vertices of a northern triangle should fall between two vertices of a southern triangle. For this I first had the machine draw a lot of such triangles, far too many for a good picture. Then, in successive tracings I found the few that appeared to look good.

The computer was little help for the surface spanning the knot beneath. The challenge of getting this object to look right led to a lengthy preoccupation with surfaces spanning knots which matured into the final picture story of this book. Perhaps it was a confidence in drawing corkscrews, inspired by innumerable spirals seen on the computer screen, that helped me draw a picture sufficiently robust to withstand reproduction in the newspaper article by Kneale [1983].

CABLE KNOT AND COMPANION. *Figure 14.*

In the same paper of Thurston, there were other knots to be drawn. The computer furnished remarkable stereographs of the knot twisting about a torus eight times as it goes around the long way thrice. But none of the views worked on its own except this one, which depends on conventional shading for its effect. Only the pen, and perusal of Escher's graphics, helped in designing the knot pair at the bottom. It was essential to depict a knot within a knot. Windows or cross sections would have detracted from the continuity. I had no choice but to try my hand at a transparent outer knot, just as Thurston had suggested in the manuscript.

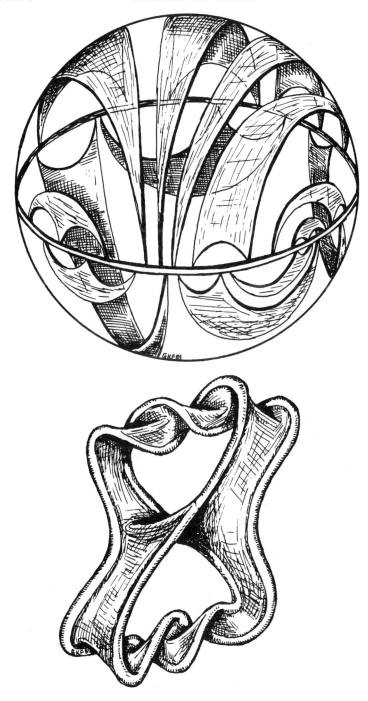

Figure 13   Lamination and Eight Knot

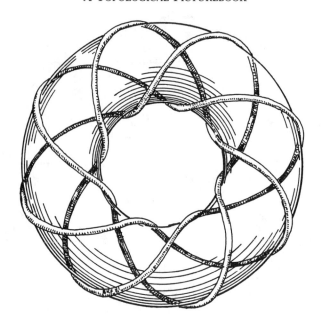

Figure 14 CABLE KNOT AND COMPANION

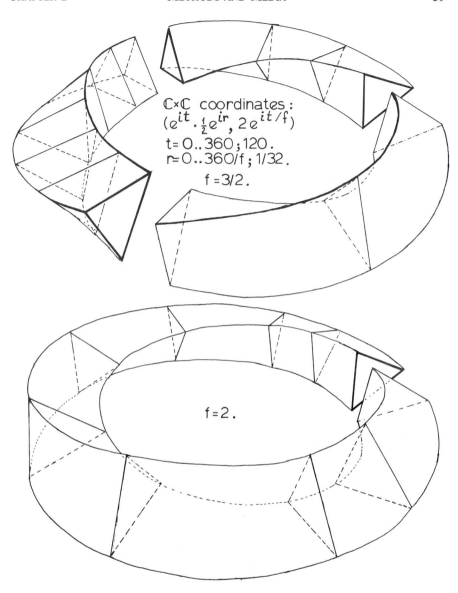

Cabling Template

Figure 15

A standard way to produce 3-dimensional manifolds is to identify the faces of a solid polyhedron in a particular way. This idea goes back to Poincaré, and in Chapter 6 we shall look at a simple example of this, where the geometry on the building block is our familiar Euclidean one. In Thurston's world, the polyhedron usually has a non-Euclidean geometry on it. When the geometry is hyperbolic, no really truthful pictures are possible. What one can hope to do is to suggest the negative curvature in a variety of ways. In a number of instances, I had to depict a process that glued one triangular face of a polyhedron to another, which a twist in between. It is remarkably difficult to imagine a turning triangle sweeping out a solid track in space. I finally made the computer produce an underlay for these two pictures of 3-stranded cable knots. These became the model for the following two free-hand illustrations for Thurston [1986]. They depict acylindrical 3-manifolds that do not have a unique hyperbolic structure. You will have the opportunity to explore the geometry of these strange 3-dimensional spaces in Chapter 4 and Chapter 8. The next chapter is about the geometry of perspective, with which we are all familiar from our comic-book days.

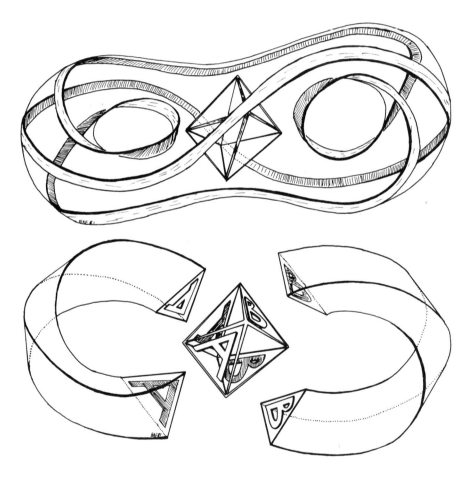

Figure 16 OCTAHEDRAL HYPERBOLIC MANIFOLD

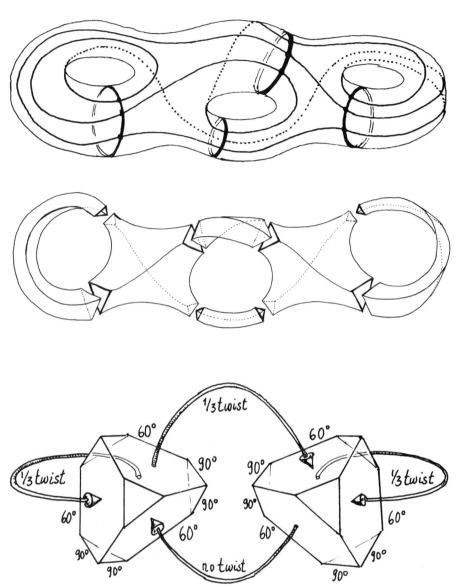

Figure 17 TETRAHEDRAL HYPERBOLIC MANIFOLD

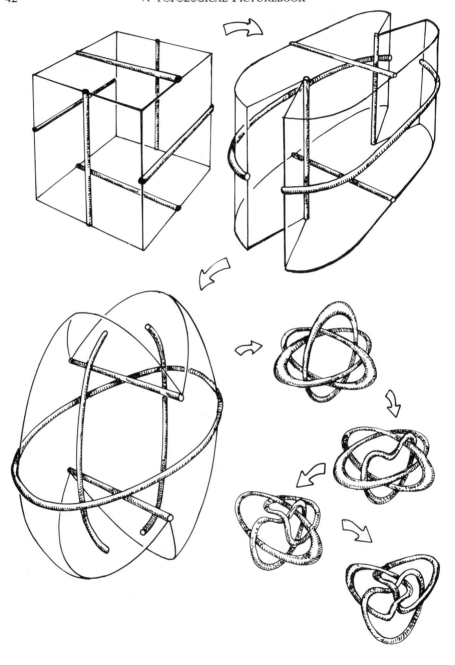

Figure 18   BORROMEAN ORBIFOLD

3
PICTURES IN PERSPECTIVE

Perspective is the simplest and most direct way of creating the illusion of depth in a picture of spatially extended objects. The more or less correctly placed vanishing points of parallel lines, the estimated regression of evenly spaced points on a line, the elliptically compressed circles: all these tricks of perspective do more than merely please the eye. They help the viewer guess correctly where the artist meant to place things relative to each other. For example, even a modest amount of perspective convergence prevents you from mistaking a three-dimensional picture for a two-dimensional diagram.

Here I am speaking of *linear perspective* as opposed to *aerial perspective*. The latter relies more on shading and shadows to give the illusion of depth. Pictures in linear perspective also differ from drawings based on an *affine projection*, where points in space are cast to the picture plane along mutually parallel lines, the *projectors*. In this case there is no convergence: parallel lines in space remain parallel in the picture. This is often called a "parallel projection" but the term should be avoided in the company of artists and draftsmen. For them a parallel projection is a perspective projection in which the principal plane of the pictured object is parallel to the picture plane. An affine projection whose projectors are perpendicular to the picture plane is called *orthographic*, otherwise it is an *oblique projection*. This is the original meaning of "orthographic" although today this term usually refers to a composite of two or more "parallel" projections at right angles to each other: the *plan* and *elevation* of a building, for example. We shall have a brief look at affine and aerial "perspective" at the end of the chapter. A clear and complete presentation of all these matters is in the classic text on engineering drawing by Thomas French and Charles Vierck [1911].

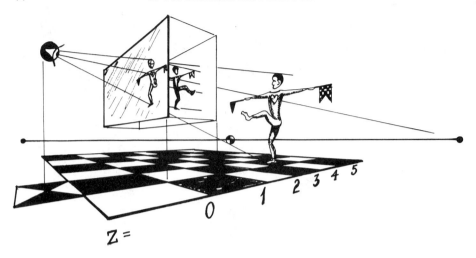

LINEAR PERSPECTIVE

Figure 1

The principle for drawing in perspective is deceptively simple. Painting from life, the artist puts a dot on the canvas where there is a real point to be seen from a fixed position of the eye. Geometrically, this is the point where the line of sight between eye and object crosses the picture plane. Since Albrecht Dürer and the Rennaissance, there has been a steady evolution of ingenious mechanical devices for facilitating this projection. Today, the photographic camera takes effortless and perfectly perspective pictures of actual scenes: landscapes, buildings, rooms and models. Inexpensive, graphics-oriented microcomputers perform this task for virtual shapes whose points have only mathematical existence. Choose Cartesian coordinates for which the eye, at $(0,0, -\delta)$, is located a *focal distance* δ behind the xy-picture plane. The perspective image, $(X, Y, 0)$, of the point (x,y,z) is given by

$$X = x Z$$
$$Y = y Z$$
$$\text{where } Z = \frac{\delta}{(\delta + z)}.$$

In this formulation we have compressed the object half space, where $z > 0$, into the region where $0 < Z < 1$. It is a reflected relief. For example, the eye 1(11) at $z = -1\frac{1}{2}$ sees the picture 1(12) at $z = 0$ of the object 1(14) at $z = 3$ whose relief image 1(13) at $z = \frac{1}{3}$ is sandwiched between the picture plane and its parallel at $z = 1$. The relief transformation is invertible:

$$x = X/Z$$
$$y = Y/Z$$
$$z = (1 - Z)\delta/Z,$$

and it could be argued that the recognition of a picture involves mental inferences about the perspective scaling parameter Z.

Before photography, and especially before computer graphics, the designer had no choice but to use the methods of classical descriptive geometry to lay out perspective pictures. Even so, considering the tedium of building physical models to be photographed, of parametrizing irregularly curved surfaces or of entering the myriads of data points needed by the computer, traditionally rendered drawings are still preferred in contemporary design. For convenience and speed, the perspective sketch is also the best choice for descriptive topologists.

Instruction for drawing in perspective can be found in primers for students of architecture, the fine arts, and industrial design (in order of increasing utility). The architect rearranges familiar objects in a familiar universe; the artist mostly draws from life. Only the designer faces problems similar to those of the topologist: to (re)create the object in the act of drawing it. Industrial and topological designers both strive to transmit visual information with minimal ambiguity and maximal economy of style. The difference comes in the nature of the pictured objects themselves. A teakettle or an automobile is, after all, part of the viewer's everyday experience. A picture of a new object of this kind serves chiefly to "reset the parameters" of an idea already in the imagination. Most pictures drawn by the topologist are of unfamiliar objects. Only rarely do they correspond to any physical reality. While a Möbius band is easily made from a strip of paper, how to sew a disc to its boundary is not a skill taught in the schools.

Hand-drawn perspective plays a special role in descriptive topology. Since perspective cubes, cylinders and cones used for framing a drawing can be copied from the computer screen, it is not absolutely essential. However, it does help the topologist to visualize and then design the object in the first place, by whatever means the finished product will be realized. Finally, I have found that teaching projective geometry, espectially to artists and designers, is best done in terms of perspective drawing.

Here, however, I shall assume that you already know the mathematical rudiments of Poncelet's ingenious creation and use it to explain perspective. Recall that *projective space* consists of ordinary 3-space augmented by an *ideal point* for each class of mutually parallel lines. An ideal point is joined to a real point by its representative through that point. Two ideal points, represented by two non-parallel real lines, are joined by an *ideal line* corresponding to the class of mutually parallel planes determined by the two lines. These ideal lines and points reside on an *ideal plane*, also thought of as the plane-at-infinity. In analogy to the *projective plane*, whose points correspond to antipodal pairs of points on a sphere under central projection, projective space has a Euclidean model in 4-space, consisting of antipodally identified objects on the 3-sphere.

That two parallel lines "meet at infinity" and that two parallel planes have a common "infinite line" has the following perspective interpretation.

As the object point at the end of the sightline moves away from the observer along a straight line in space, its trace on the picture plane moves along a line segment until it stops at the *vanishing point* of the object line, which is now parallel to the sight line. For this reason, the vanishing point remains fixed as long as the object line moves parallel to itself in space. As long as the object line rotates in a plane, its vanishing point moves along a line, called the *horizon line* for the plane of rotation. Thus the picture plane comes into 1:1 correspondence with the ideal plane under a projection along lines of sight. The vanishing points of object lines parallel to the picture plane are the ideal points for the picture plane. Its ideal line is the horizon for planes parallel to it.

HORIZON AND ZENITH. *Figure 2.*

Let me now take you through some easily remembered perspective tutorials. Even if accurate execution of these drafting procedures is not always practical, I find the approximation to them useful for sketching a picture freehand. Their demonstration furnishes concrete examples of very beautiful theorems in geometry. They cluster about the following three construction problems:

1. Given a horizon line, find the corresponding *zenith point*, which is the vanishing point of the lines perpendicular to a plane with that horizon. This is a special case of perspective measurement of angles and distances.
2. Complete the drawing of a cube starting from a segment and horizon line in the picture plane which represent an edge and a face plane of the cube. In a sense, all traditional perspective methods begin with a solution to this problem. My version resulted from a study of the monograph on perspective drawing by Jay Doblin [1956] and the detection of geometrical errors made for aesthetic reasons. I am grateful to Ben Halpern for convincing me of their presence.
3. Given a square and its perpendicular in a picture, discover and use its perspective information to draw other common figures, such as circular wheels with perpendicular axles.

To solve the first problem, analyze the situation from a third-party view, 2(11). A right triangle has been placed between picture and observer. The hypotenuse B-Z is in the picture plane PP; the apex A of the triangle is at the eye of the perspectivist; and one leg A-B is in the sight plane SP to the given horizon line $h\ell$. The altitude A-C of the right triangle connects the eye to the nearest point C on PP, which I call the *focal center* or center of vision; and its length is the *focal distance* δ. The interior of the circle centered at C and radius δ in the picture plane is the *focal disc*. It corresponds to a 90° visual cone from the eye point. Obviously, the other leg A-Z of the right triangle is a sightline perpendicular to SP, and hence Z is the desired zenith.

Chapter 3 — Pictures in Perspective — 47

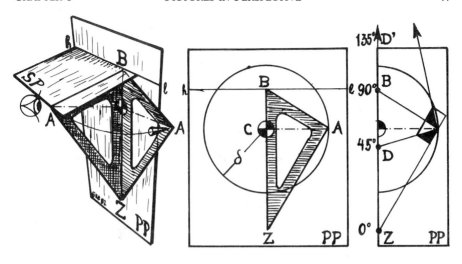

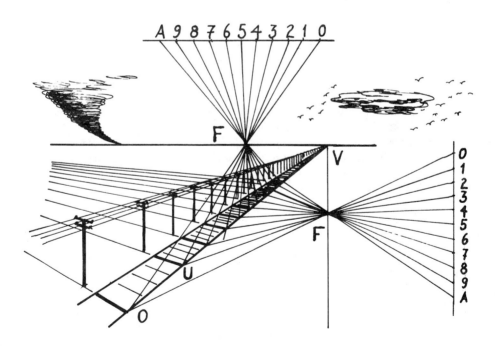

Figure 2   Horizon and Zenith

To synthesize this construction in a picture plane, rotate the triangle about its hypotenuse, 2(12). For now I shall assume that the focal disc is given, and tell you later how to determine it from other clues in a given picture. For any horizon line $h\ell$, erect a right triangle with one acute angle at B, the foot of the perpendicular from C to $h\ell$, and apex A on the focal circle. The other end of the hypotenuse is the zenith point of $h\ell$. This way, a line sticking out of an object plane at any angle θ can be drawn towards the θ-vanishing point on the axis B-Z. Most important are the points D for $\theta = 45°$ and D' for $\theta = 135°$, because these are the vanishing points of the diagonals of a square whose edges vanish at B and Z.

Geometers may recognize that D' is the harmonic conjugate of D relative to Z-B. Or, what comes to the same thing, the cross-ratio $CR(D';Z,D,B) = -1$. Since it is the "perspectivist's ruler", a brief digression on the *cross-ratio function* is appropriate. For three points, O-U-V, on a line directed towards V, the function

$$CR(X;O,U,V) = \frac{\langle X - O \rangle}{\langle X - V \rangle} \frac{\langle U - V \rangle}{\langle U - O \rangle}$$

assigns numbers to all points X on the line, once you give terms like $\langle X - Q \rangle$ the value of the signed distance from Q to X. Note that the function has values $0,1,\infty$ when $X = O,U,V$. The mnemonic for the cross-ratio is: O for "origin", U for "unit" and V for "vanishing point". The cross-ratio is negative at points on the observer's side of the origin and beyond the horizon. When V is chosen to be the ideal point on a line, you should cancel the two infinite terms and thereby obtain an evenly spaced scale on the line with unit $\langle U - O \rangle$.

The real utility of the cross-ratio derives from its projective invariance. Suppose that on two lines the O-U-V triples line up so that the three lines through corresponding points cross at the same place, call it their *focus F*. The respective cross-ratio functions are equal at corresponding points X. Given, for example 2(21), the first two telephone poles O and U along the prairie railroad vanishing at V, you can propagate the poles as follows. Choose any convenient focus F and point A not collinear with V and F. On the parallel to V-F through A the scale corresponding to O-U-V via F is evenly spaced because V is sent to the ideal point. You should convince yourself of the invariance by a rewarding exercise in trigonometry. Use the law of sines to show that the terms such as $\langle X - Q \rangle$ in the cross-ratio may be replaced by the sine of the directed angle QFX. When F is an ideal point, the correspondence is effected by parallel lines. Hence $\langle X - Q \rangle$ may also be replaced by the difference of the x-coordinates, insofar as the line is not parallel to the x-axis. This simplifies machine computation of cross-ratios.

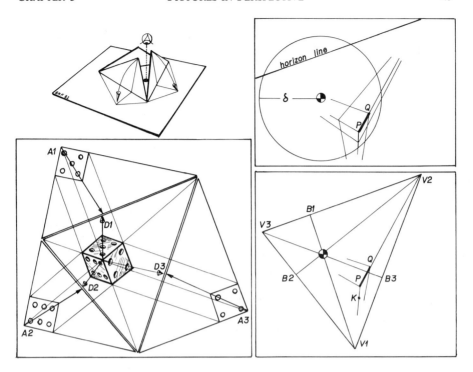

Recipe for a Cube

Figure 3

The second problem, completing an edge and face to a cube, can also be understood in terms of tilting a right triangle about its hypotenuse into the picture plane. Suppose you are looking at the corner of a cube in space so that three faces are visible. Imagine these faces extended away from the corner, and then translate this triangular cone to your eye. This *sight cone* of the cube crosses the picture plane on the *framing triangle* whose vertices are the three vanishing points of the edges of the cube. The three edges of the framing triangle are the horizon lines for the faces of the cube. Now tilt each of the right triangles which the picture plane cuts from the sight cone, into the picture plane. In 3(11) you see the sight cone opening up like a barnacle, as its faces tilt to the outside of the framing triangle. You could also collapse the sight cone to the inside of the framing triangle.

A moment of geometrical reflection reveals a remarkable set of coincidences. If you prolong the three altitudes into the framing triangle, 3(21), then they pass through the focal center on their way to the zeniths at the vertices. The point where the altitudes of a triangle meet is the *orthocenter*. In other words, the orthocenter of an acute triangle, 3(22), is also the focal center of a perspective view for which opposite vertex-edge pairs are the zenith-horizon pairs of mutually orthogonal planes. Figure 3(21) is an orthographic drawing of a die placed into the apex of the sight cone. The position of the three visible faces after they have been tilted into the picture plane are drawn on the outside of the framing triangle. Note how the three diagonal vanishing points can be located on the framing triangle by prolonging the diagonals of the three little squares from AI to DI, $I = 1,2,3$.

With all this understood, you are ready for a recipe that completes the cube, 4(11), given the picture plane location of the nearest vertex P, an edge P-Q, and the horizon line for one face containing this edge, 3(12). Construct the zenith $V1$ for the given horizon, 3(22), using the focal disc as above. Prolong the given edge P-Q till it crosses the horizon line at vanishing point $V2$. Prolong the perpendicular, $B3$-C, from the edge $V1$-$V2$ through the focal center to find the third vertex, $V3$, of the framing triangle. The four lines, P-$V2$, P-$V3$, Q-$V1$ and Q-$V3$, depict the corner of an infinite slab which is one edge-unit thick. To carve a cube from this, we need a second corner, K nearest $V1$ for example. But K is where P-$V1$ crosses the prolongation of the line from $V2$ to that corner on the cube where the line Q-$V1$ crosses the diagonal P-$D3$. That is what diagonal vanishing points are very good for. This produces an infinite beam, one edge-length square. As soon as you have found one more corner, say with the aid of a second diagonal vanishing point, the remaining constructions become obvious.

CUBE IN 3-2-1-POINT PERSPECTIVE. *Figure 4.*

Drawing 4(11) summarizes this recipe for the case when the framing triangle is in the finite part of the picture plane. To check that you understood the recipe you should label 4(11) to match 3(21) and 3(22). Then label 4(21) and 4(22) analogically. When one plane of the object cube is perpendicular to the picture plane, 4(21), then the its horizon line passes through the focal center and its zenith is an ideal point of the picture plane. The cube is said to be drawn in *two-point perspective*, because only two of the three vanishing points are finite. As before, prolong P-Q to find $V2$. Since the right triangle shown was tilted 90°, it can be reconstructed by erecting a line $A1$-C one focal distance long and perpendicular to the horizon line. Then complete the right angle $V2$-$A1$-$V3$, and locate the diagonal vanishing point $D1$ as before.

There is a quicker way of finding the diagonal vanishing points $D3$ and $D2$. Note that the face C-$V2$-$V1$ of the sight cone is the rectangular strip $V1$-C-$V2$-$V1$, tilted into the picture plane. Since the same edge on the sight cone tilts to $A3$-$V2$ and to $A1$-$V2$, both diagonal vanishing points $D3$ and $D'3$ are on the circle $V2$ through $A1$ and $A3$.

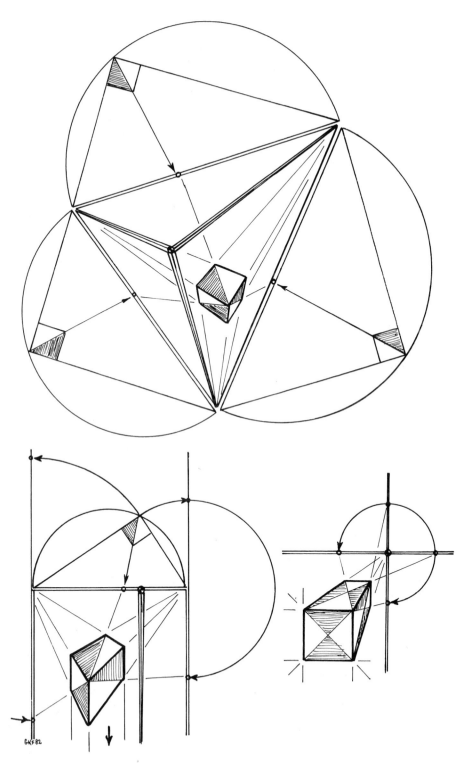

Figure 4   Cubes in 3-2-1 Point Perspective

For a cube in *one-point perspective*, a face is parallel to the picture plane. Hence one direction vanishes at the focal center, sending the other two vanishing points to infinity. Since one edge of the sight cone is perpendicular to the picture plane, the focal circle passes through all four diagonal vanishing points located on the two perpendicular horizon lines. The third horizon is the ideal line of the picture plane. Traditional tutorials exploit these special properties of 1-point and 2-point perspective to devise much more efficient but also more complicated drawing strategies than mine.

Looking for a Cube. *Figure 5*.

To insure that all finite vanishing points are on the figures, I have chosen small focal distances. The 3-point cube in 4(11) looks fine, while both of the other two cubes look distorted. Despite the acute corner on 4(21) and the square face on 4(22), these are drawn in true perspective. An unusually nearsighted viewer could see them correctly from the eye point a few centimeters above the focal center. If you erect a cube in 4(11) but with starting edge *P-Q* outside the frame, it would look equally distorted. 5(11) shows a cardboard device I use to convince skeptics of this. It shows a part of a sight cone, with a peephole in its vertex. The opaque cone serves to mask the surrounding visual cues as you look at a 2-point perspective cube well outside the focal disc. Since the picture can be quite far from the peephole, it isn't necessary to make the focal distance, and hence the model, very large at all.

Let me now apply the recipe by reconstructing a perspective in which an arbitrary convex quadrilateral looks like a square. The idea stems from Ben Halpern. Prolong opposite edges, 5(21), to locate the vanishing points and horizon 1-2. Extend the diagonal to its vanishing point 3 on the horizon. Bisect 1-2 at 4 and drop a perpendicular radius 4-5 to the circle 152. Prolong 5-3 to 6 where it bisects Thales' triangle. (This peripheral angle is 45° because it "sees" 90° of the perimeter.)

Therefore, the focal center, *C*, must be somewhere on the altitude 6 − 7 or its reflection, 7-6′, in the diameter. Since the focal center will be the geometric mean of the section 6-*C*-6′ it is longest in 2-point perspective: *C* is on the diameter, as drawn. To draw parallel squares, 5(22), connect a point 1 to vanishing points 2,3 and 4. Choose point 5 on the diagonal and complete the square 6,7. The quadrilateral looks increasingly less "square" as it flees the focal disc. To erect the cube 5(31) find the diagonal vanishing points 1,2, on the horizon whose zenith is the opposite vanishing point. Connect 2-3 and use its crossing 5 with one vertical edge to find 6, then 7 on the other two vertical edges. I leave the 3-point cubes to you.

If the initial "square" is a trapezoid, 5(32), use the recipe for a 1-point perspective. However, when both initial vanishing points are at infinity, that

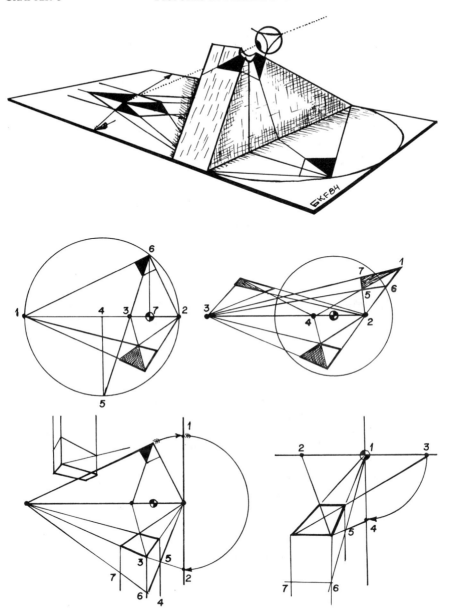

Figure 5 LOOKING FOR A CUBE

is, when the initial quadrilateral is a parallelogram, you must seek an affine projection in which to "see" the square. The one exception to this is hidden in 5(32).

DIAGONALS. *Figure 6.*

The scale at which most of us are likely to draw has a focal distance of 20 to 30 centimeters. This places most vanishing points outside the drawing surface. Thus there is a choice between technically accurate but exaggerated perspective, and more natural approximations using estimated locations of points and lines. Here is a simple procedure for drawing a third line perspectively parallel to two given lines when the vanishing point V is inaccessible. Given two lines $\ell1$ and $\ell2$, converging at V, and a point O, construct a line through O towards V as follows. Erect three parallel lines crossing the given lines so that an outside one goes through O. For the case that O is inside the angle, 6(11), I have used the paper margin opposite V as a parallel. Locate 5 by prolonging the line from O through the diagonal point of quadrilateral 1234. Likewise, find 8, using 5 and 2467. Since the lengths of the segments are in these proportions:

$$|O1| : |13| = |25| : |54| = |68| : |87|,$$

the prolongation of the line from 8 to O goes through V. For the case that O is outside the angle, 6(12), use the V-ward margin as parallel, find 2 on O-1, prolong 3-2 till 4, find 6 on 4-5, and prolong 1-6 till 7. The analogous proportions predict that the prolongation of O-7 will pass through V.

The subdivision and multiplication of squares and cubes in perspective, by means of the diagonal point, is a convenient way of estimating metric points without elaborate perspective measuring machinery. For example, 6(21), shows how to construct a checkerboard from an initial square *ABCD* without using cross-ratios. Prolong two sides of the square and draw the centerline, 1-2, of this "railway track" through the diagonal point 1 and the estimated vanishing point. Prolong *A*-3 to 4, and build the tie 4-5 by estimating the vanishing point of the other two direction. This duplicates the square the first time. Repeat forever.

In space, 6(31), diagonalize the three visible faces of a given perspective cube, *ABCDEF*, to find centers 1,2,3. Estimate centerlines, as before, to find edge bisectors, such as 4,5,6. This suffices to subtract an eighth of the cube, 6(32). Prolong 1-3 to the base centerline to find 7, in order to build the addition.

CHAPTER 3 PICTURES IN PERSPECTIVE 55

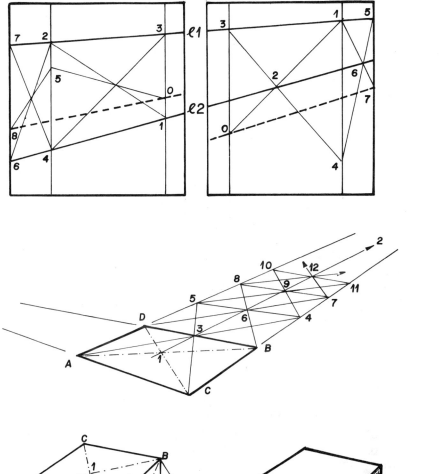

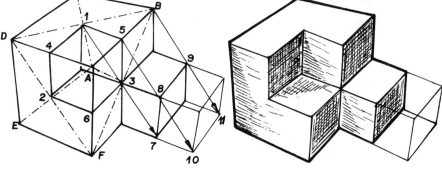

Figure 6 DIAGONALS

WHEEL AND AXLE. *Figure 7*.

The second diagonal subdivision of a perspective square into 16 tiles, 7(11), provides a practical way of estimating an inscribed circle. 7(12) shows the corresponding metric square. From each of the four corners, draw two rectangular diagonals towards the quarter marks on the far sides. The points where these construction lines exit the corner tile are just outside the circle. These eight points are the symmetric reflections of, say (½,⅞), and

$$(4/8)^2 + (7/8)^2 = (65/64) \sim (8/8)^2.$$

You now have 12 points, including the perspective bisectors of the edges, through which you can interpolate an ellipse freehand or with a french curve.

A fair estimate for the axle of a disc is the line through the center of the perspective square and in the direction of the minor axis of the ellipse depicting the inscribed circle. This is truer, the closer the figure is to the focal center; for there the perspective view is nearly orthographic. But at high noon, 7(13), the axle of this gyroscope projects directly to the minor axis of the rotor's elliptical shadow on the ground. As the figure moves away from the focal center, the apparent distortion, as seen from an improper eye point, increases. In 7(21) you see six parallel discs. The two uppermost are translates of the central disc in the two orthogonal directions and in its own plane. This plane, incidentally, is tilted about 45° relative to the picture plane, as you can check by noting that the zenith is (nearly) on the focal circle. The nearer two circles in this plane have diameters ½ and ¾ respectively. The bottom disc is a translate along the axle of the central disc. You can decrease the distortions by looking through a masked peephole (punch a hole in a file card) at a distance δ above the focal center.

There are many other practical ways of drawing feehand in perspective, to be found in Jay Doblin's monograph [1956]. He develops an ingenious method for minimizing the marginal distortion that comes from looking at the picture from much further away than the focal distance. To draw the checkerboard, for example, he fixes the two vanishing points on the horizon line, but varies the diagonal vanishing point for each square. This places the focal center of each square on a vertical line through the nearest corner. Doblin's roving focal center is reminiscent of panoramic photography and 15th century painting in that it respects the viewer's habit of scrutinizing a detail by shifting the picture directly before the eye.

CHAPTER 3 PICTURES IN PERSPECTIVE 57

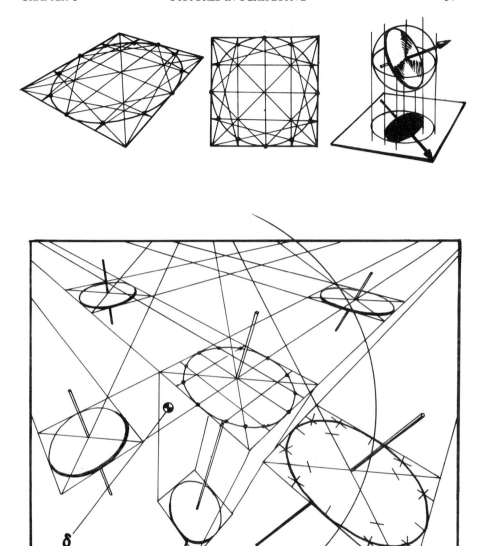

Figure 7 WHEEL AND AXLE

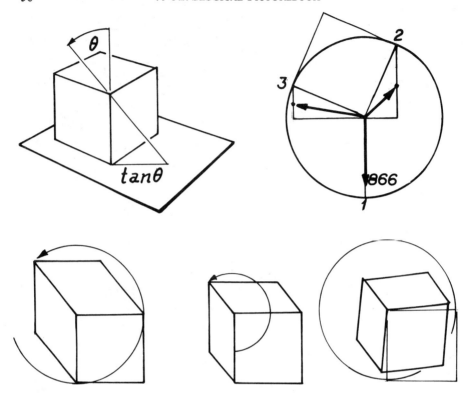

AXONOMETRICS

Figure 8

A mathematician, about to draw a 3-dimensional picture, usually begins with a system of coordinate axes. Usually, he draws the third axis in some direction and some distance from a corner of the right angle. Complete the right angle to square for easier reference. The position in the picture plane of a point (x,y,u) in space is determined by plotting (x,y) in the grid of the square, and moving this point parallel to the third axis a distance u times the length of that axis. That this invariably produces an oblique projection can been understood as follows. Imagine a cube resting on the square in the picture plane, 8(11), then the third axis is the shadow of an edge normal to the plane. If the projectors are declined an angle θ from normal, then the shadow has ratio of $\tan(\theta°)$ to the edge unit.

In the *cavalier projection* all axes in the picture plane have the same length and two are at right angles, 8(21). Hence the projectors are declined 45°. To reduce the distortion inherent in this very oblique projection, the *cabinet projection* takes the third axis half as long as the square's edge, 8(22). It corresponds to an angle of 26°34′ from normal for the projectors. By widening the right angle of the cabinet projection a bit, 8(23), we approach an orthographic projection of a cube.

Here is an instructive way to analyze the conditions for a given set of axes in the picture plane to fit the orthographic projection of some cube in space. Recall that the set of 3 by 3 matrices P whose transpose is equal to their inverse, $P^T = P^{-1}$, form the *orthogonal group*. Such a matrix is also called *orthonormal* because $PP^T = I$ says that the three column vectors are mutually perpendicular unit vectors. Generally, we prefer a right handed system, so that $\det(P) = +1$ and the third column vector is the cross product of the first two. These matrices constitute the group $SO(3)$ of "special" or *sense-preserving orthogonal* transformations in 3-space. Thus, the three row vectors of P are also an "orthonormal" because they are the columns of its transpose. Now suppose the 2 by 3 matrix

$$M = \begin{vmatrix} x_1 & x_2 & x_3 \\ y_1 & y_2 & y_3 \end{vmatrix}$$

expresses your choice of axes in the picture plane. Then the axes are orthographic if and only if

$$x_1^2 + x_2^2 + x_3^2 = 1 \tag{1}$$
$$y_1^2 + y_2^2 + y_3^2 = 1 \tag{2}$$
$$x_1 y_1 + x_2 y_2 + x_3 y_3 = 0. \tag{3}$$

To obtain a canonical form for M, rotate the picture plane until $y_1 = 0$ and x_1 is the cosine, C_1, of the angle θ_1 that edge number 1 of the cube makes with the picture plane. Equation (2) then says that $y_2 = S_2$ and $y_3 = C_2$ for some suitable angle θ_2. Equation (3) requires that (x_2, x_3) be proportional to $(C_2, -S_2)$. Equation (1) now forces M to be

$$\begin{vmatrix} C_1 & S_1 C_1 & -S_1 S_2 \\ 0 & S_2 & C_2 \end{vmatrix}$$

A geometrical interpretation of this canonical form is shown in 8(12). Choose direction 1 for the first axis on the unit circle and locate a square 2-3 turned an angle θ_2 from 1. (Here, $\theta_2 = 155°$ for no particular reason.) If the first axis is shortened by a factor C_1, then the other two axes must suffer a shear by a factor of S_1 towards the perpendicular to direction 1. In example 8(12), $\theta_1 = 30°$, so that $S_1 = \frac{1}{2}$, and the orthographic correction is as shown. On the other hand, since 8(23) began as a cabinet projection, $C_1 = \frac{1}{2}$ by definition. Hence the shear, which makes the right angle obtuse, shortens those sides by only 13 percent. Thus the cabinet projection really is close to orthographic.

On the other hand, if three *a priori* directions are to be preserved at all cost, then the lengths may be adjusted according to the following limitations. Subtract constraint (2) from (1), double constraint (3) and switch to complex arithmetic, $z = x + iy$. The constraint now says that the complex squares of three axes "balance" over the origin:

$$z_1^2 + z_2^2 + z_3^2 = 0 .$$

An orthographic projection is *isometric* if all three axes have equal length. It follows from this quadratic equation that there is only one isometric projection. Its axes are 120° apart, and the contour of the cube is a regular hexagon. A less obvious though equally direct consequence is the *obtuse angle condition* for three given directions to fit some orthographic projection. We may assume that the given directions span the plane, i.e. that we are drawing the nearest corner of a cube. Assume that the least of the three angles is from z_1 to z_2 and that $z_1 = 1$. Then the spanning hypothesis puts z_3 in the third quadrant. Hence its square is in the upper half plane. If z_2 were in the first quadrant then all three squares would share a half plane and the quadratic constraint would fail. If, on the other hand, all three angles are obtuse, then z_2 is in the second quadrant, its square is in the lower half plane, and appropriate positive weights make things balance over the origin.

No restrictions apply to the choice of axes in the picture plane if you don't mind an ugly, oblique projection. You can discover the tilt of the projectors algebraically as follows. For any 2 by 3 matrix M of rank 2, there is an orthogonal matrix P and a diagonal matix Q such that

$$I = Q P M M^T P^T Q^T .$$

The columns of the rotation P are the unit length eigenvectors of the positive definite symmetric matrix MM^T. The square roots, $q_1 < q_2$, of the eigenvalues appear in the scaling matrix Q. The projectors are tilted an angle of $\arccos(q_1/q_2)$ from the normal in the direction of the larger eigenvalue of MM^T. You should exercise your skill with matrices by checking this rule for the first case we considered, 8(11). Let

$$M = \begin{vmatrix} rC & 1 & 0 \\ rS & 0 & 1 \end{vmatrix}$$

represent the mathematician's choice of axes. Show that $r = \tan(\theta)$, where θ is the declination angle of the projectors in the direction $\langle C, S \rangle$.

CHIAROSCURO. *Figure 9.*

The geometrical theory of light and shade has its origin in two eighteenth century treatises. Johann Lambert's "Photometria, sive de mensura et gradibus luminis, colorum at umbrae" appeared in 1760 and Pierre Bouguer's 1729 "Traité d'optique sur la gradation de la lumière" was published in the

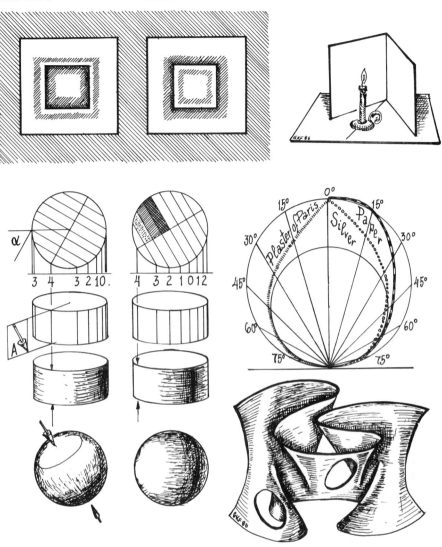

Figure 9 CHIAROSCURO

same year, see Rohn-Papperitz [1906, p.193]. Here I shall not be interested in the problem of depicting *shadows* cast by one object on another. The shape of a shadow still obeys the objective laws of linear perspective, see Krylov-Lobandievsky-Men [1968]. The question here is which part of an object should be drawn just how much darker than another. The gradation of apparent brightness on a curved surface depends ultimately also on the physiology of the human eye, a factor not well understood until the work of Ernst Mach [1886].

The didactic purpose for shading pictures in this book at all, and the limitations of the chosen media, will make my review of the theory very brief. Referring to the perception of a pictorial illusion, such as the one in the next chapter, Scott Kim [1978, p.210] calls "Vision ... a learned skill. Translating 2-d images to 3-d objects requires the aid of an acquired vocabulary of expected patterns." Accordingly, the shading technique I use makes no pretense of accuracy or realism. It merely encodes positional information and helps distinguish rounded contours from sharp borders. It is based on a few optical principles.

The first principle is the *inverse square law*, which accounts for the observation that light from 4 candles at 2 meters seems as bright from 1 candle at 1 meter. Therefore, all things being equal, when two surfaces overlap, the one further away should be darker than the nearer one. However, it is impractical to use as many shades of grey as a rigid application of this principle would require. I rely on the following optical illusion instead. The *Mach effect* produces an apparent bright line parallel to a white field in an adjacent dark field and vice versa. This is the result of the lateral inhibition of retinal cells. You will find a convincingly illustrated explanation of this (and other illusions) in Colin Blakemore's article in Gregory-Gombrich [1973]. In order for one of two adjacent fields of equal grey value to appear globally brighter than the other, Blackmore produces a sharp gradient locally along the border. Here is an application of this principle. It is easier to accept the suggestion that the square on the left, 9(11), has a square hole, and that there is a little square hovering in front of the big square on the right, 9(12), than the contrary. Yet the two details only differ in the succession of grey-tone across the smaller square.

Rules for shading, such as *Lambert's cosine law*, should account for the fact that luminous spheres, such as the sun, as well as illuminated spheres, such as the moon, seem uniformly bright. Both look like discs. Consider, for the moment, the case of the moon. If she were a perfect mirror, we would see the sun reflected in exactly one point, the so called *brilliant point* (Glanzpunkt). Evidently, the surface irregularities scatter the non-absorbed light rays in all directions. Now the area of the retinal image of a surface element seen from an angle β from its normal direction is reduced by a factor of $\cos(\beta)$. Let us suppose, with Lambert, that the intensity of light emitted in the direction β is also reduced by the factor $\cos(\beta)$ from that emitted in the normal direction. Then the retina receives the same illumination per unit area from every spot on a luminous sphere. It will appear to be flat. On the other hand, the energy per unit area received when a surface is illuminated by a single, distant source (the sun) is proportional to the cosine of the angle of incidence. Thus the gradation of apparent brightness on a curved surface obeying Lambert's law is proportional to the cosine of the angle of incidence, and is independent of the viewing angle. The full moon illusion must have a different explanation.

One of Lambert's experiments, as reported by Wiener [1884, p.394] is suggested by 9(13). The two faces of a paper dihedral, equally illuminated by a source on the bisector, look equally bright, and project equally bright images through a lens on a screen, no matter what the viewing angle β is. Curves on a surface of equal illumination are called *isophotes*. On a cylindrical surface the isophotes are generating straight lines, 9(31) and 9(32); on a sphere they are circles perpendicular to the axis through the brilliant point. In 9(21)-9(51) this axis is inclined an angle A from the horizontal (base) plane, and 60° in the base plane from the viewer. Let α measure the angle of incidence of the plan projection (top view), 9(21). Then the cosine of the angle of incidence equals $\cos(A)\cos(\alpha)$ and the plan projection can be used to map the isophotes, 9(31). Divide the diameter through the brilliant point into 8 equal sections, 4 for the illuminated half, 4 for the half in the shadow. Then the illumination on an isophote labeled 3 is ¾ as bright as on the isophote labeled 4. In 9(22)-9(52) the light source has moved.

It is customary to associate a *backlight* to each light source. It is parallel, but opposite the source, and of reduced intensity. The backlight for 9(51), little arrow, is not significant, but it makes 9(52) seem much more 3-dimensional. Try it on a xerox of this illustration by blackening the shadow part of 9(42) and 9(52).

As you will have noticed, I prefer to shade a picture by means of grids of parallel curves on the surface. It is faster than applying random dots and survives xeroxing better than solid grey and black fields. Two additional advantages are not so obvious. The *half-tone* is produced by a "wave-length" of 2 pen-widths. Look at 9(22). Applying a second grid of the same wave-length halves the brightness again to form the *quarter-tone*. On the other hand, continuing the half-tone as dashed lines suggests the ¾-tone. The choice of direction for the half-tone lines is also important. It establishes continuity of a sheet that is partially hidden by a nearer one, as in 9(12). In 9(11), the direction of the grid in the window does not match that of the general background. It is not clear whether you are seeing the background or a third, different level. On your xerox copy, replace the little square by one shaded in the same direction as the background to verify this illusion.

Many factors influence the observed deviations of matte surfaces from Lambert's law. The results of Bouguer's experiments are summarized in Figure 273 of Wiener [1884], redrawn in 9(23). Three handy materials, plaster, matte silver and a kind of drawing paper, were illuminated at angle zero (the sun is behind the viewer), and measured at various angles of incidence. (It is easier to understand this figure if you think of a sphere and a stationary observer.) The two circles represent Lambert's law at two levels of reflectivity. The smaller diameter is 74 percent that of the larger, which measures the illumination at the brilliant point. At an angle of 60°, for example, the secant line from the base point is half as long as the diameter because $½ = \cos(60°)$. All three materials are duller near the edge than

Lambert's law predicts. On the other hand, you might consider them brighter in the vicinity of the brilliant point, due to the fact that none are perfectly matte and do reflect some light like a mirror.

Expert drawing teachers, such as Kimon Nicolaïdes [1975] rely on practical rules which show only qualitative respect for the laws of geometrical photometry. In the lesson on drawing drapery, he summarizes these as follows. Every fold has (at least) three surfaces. The top, facing you, should remain white. The right and left sides are grey (Lambert's law). The base, from which the fold rises, is dark grey but not black (inverse square law). Black is reserved for the *undercut* (Mach's principle). Borrowing the term from sculpture, Nicolaïdes [1975, p.112ff] says : "Where the side is turned under so that it is not visible, there is an undercut under the edge of the fold....[It] is to be indicated by a black line which graduates as it moves away from the edge of the fold, giving the impression that the pencil has reached under where the fold turned under and has then come out again with lessening pressure. Similarly, the sculptor would lessen the pressure of his tool as he came out from under an undercut." The placement of these features on your drawing "should be so clear that a woodcarver could carve a piece of wood from it without any other explanation."

I have adapted Nicolaïdes' rules to the bent surfaces of descriptive topology. In 9(53) are 4 cusps and two windows. Darkening the background, as in 9(11), is too time consuming. Accordingly a fold contour is darkened, except when it becomes an undercut. There a Mach line sharpens the contrast. Inevitably, such rules of thumb come into conflict with each other, and then a descriptive topologist does best to borrow a graphical solution from the expert artist. The works of Maurits Escher are a treasury of solved problems in linear and aerial perspective. My first picture story and the next chapter is about one of the most precious of these Escher treasures.

4
THE IMPOSSIBLE TRIBAR

Few mathematicians have never meditated on Maurits Escher's magnificent "Waterfall". This print convincingly depicts an over and undershot waterwheel perpetually driven by a closed stream. The water flows from the base of the wheel down a zigzag millrace which is supported by two towers, and spills from it again, two stories above the wheel.

Penrose Tribar. *Figure 1.*

The graphic principle behind this illusion owes its formulation to the psychologist L. S. Penrose and his physicist son Roger [1958]. It consists of three square bars (prisms), oriented in three orthogonal directions; each joined to the next by a right angle; altogether extended in 3-space, 1(12). Here is a simple recipe for drawing this isometric projection of the Penrose tribar. (1) Truncate an equilateral triangle by little segments parallel to the opposite sides. (2) Draw inside parallels, (3) truncate again and (4) again draw inside parallels. (5) Imagine one of the two possible cubes at a corner, (6) extend it to an L-shaped piece and (7) copy it cyclically about the figure. The composite, 1(11), summarizes the procedure. Had you chosen the other cube, center detail, the tribar would have "turned" the other way.

Although the isometric form of the illusion is the most practical, any initial triangle will do. A different one was used for 1(21), which shows the tribar exploded into three L-shaped pieces inscribed in cubical neighborhoods. Nor is it necessary to use affine projection (orthographic or oblique) to produce an impossible figure. Penrose drew the figure with each bar converging to vanishing points, but not in global perspective. In 3-point perspective a rotationally symmetric tribar, 1(22), does not seem able to maintain a constant cross-section. At a corner, its cross-section appears to

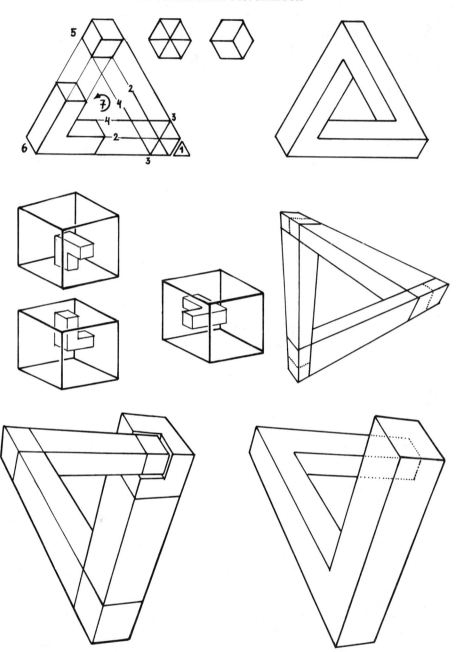

Figure 1   Penrose Tribar

Figure 2   IMPOSSIBLE QUADRILATERAL

be rectangular in a ratio of 2:1, but shrinks to a square in order to close up correctly at the next corner. Under affine projection parallel lines remain parallel and an object does not shrink as it recedes from the observer. In perspective, starting from a cubical corner, a tribar closes with a smaller cube than the one it started with. The illusion 1(31) is perspectively correct but closes incorrectly. The correct closure 1(32) shows why the tribar is really not possible in 3-space.

IMPOSSIBLE QUADRILATERAL. *Figure 2.*

Thaddeus Cowan [1974] studied a related illusion. An impossible quadrilateral is formed by truncating an isometric tribar, 2(11). The less convincing, but more symmetrical oblique projection, 2(12), can be drawn with a tripled rectangle. Add 4 diagonals and erase 26 sides of the little squares at the corners. Coding the corner formation in terms of braid group theory, Cowan found 27 recipes for assembling such figures, 21 of which are impossible. Trace the succession of faces around the (apparent) polytope 2(11), and note how its toroidal surface splits along the edges into two ribbons, 2(21) and 2(23). Each has one full twist as it writhes once, and the two borders, 2(22), link twice. As such, this makes a pretty example of Jim White's theorem that "link equals twist plus writhe" [1969]. I shall tell you more about this relation in the last picture story.

An astute analysis of the Penrose tribar, revealing the precise nature of the illusion, permits Scott Kim [1978] to construct its 4-dimensional analog. In effect, he designs a 3-dimensional polyhedron which, to a denizen of 4-space, would look like an isometric, hidden "face" picture of an impossible 4-dimensional object.

PIPELINE. *Figure 3.*

My story about the tribar differs from the others in that it begins with an act of faith that the picture displays true information about a polytope located, if not in our familiar 3-space, then in some other Euclidean manifold. Suitably interpreted, the picture contains a recipe for contructing an honest home for the tribar. The simplest, suggested to me by John Stillwell, is to first arrange three successive bars as directed by the picture, except that the terminal point is displaced horizontally $\sqrt{3}$ bar lengths from the initial point. Repetition generates an endless spiral "pipeline". The infinite cyclic group Z generated by this translation, acts on all of Euclidean 3-space R^3. The quotient R^3/Z is topologically the same as the 3-dimensional cylinder $S^1 \times R^2$. It inherits the (straight) Euclidean geometry. Think of a "wrap around" space, as the space between two mirrors, except that orientations are not reversed in the "next" room. Since every fourth bar in the pipeline is now identified with the first, the figure closes up correctly to form a tribar.

DICE. *Figure 4.*

A second recipe for constructing a 3-manifold is deciphered from the Penrose illusion as follows. Isolate each corner in a cubical neighborhood, like 1(21). As you glance from one position of the cube to the next in cyclic order, it rotates 120° about the diagonal joining the nearest corner of the cube to the furthest. In the isometric view, 4(11), this axis is perpendicular to the picture plane. Label the faces of each cube like those of a die so that opposite face labels add up to seven. The three ways the cubes can slide together pairwise, suggest gluing faces 6 and 3. To see how to identify the remaining twelve faces pairwise, imagine the die repeated infinitely in all directions. This forms a cubical lattice, like 6(11), but projected isometrically, as in 6(21) and 4(12).

Figure 3 PIPELINE

I have labeled the spots on the dice black, white and grey so that the three positions in 4(11) can be referred to by their color. From 4(12) you can read the gluing scheme of the white die's visible faces. If you trust your sense of symmetry, you can obtain the others by permuting the color labels cyclically, $W \to B \to G \to W$. This is how the faces fit together:

$$W1 - B5 \quad B1 - G5 \quad G1 - W5$$
$$W2 - B4 \quad B2 - G4 \quad G2 - W4$$
$$W3 - B6 \quad B3 - G6 \quad G3 - W6$$

To discover how the edges and vertices fit together, slide the dice together as dictated by the illusion. Label a directed edge by the dihedral angle, and a vertex by the solid angle formed by the adjacent faces. Thus, edge $W12$ is directed from vertex $W123$ to $W142$, and $W21$ has the opposite direction. In 4(12) you see how the edge W12 and its tail $W123$ tour their equivalence classes as indicated by the initial three, respectively all seven, arrows. The last vertex, $W456$, is diametrically opposite the first on the white cube in the center.

$$W12 - B35 - G46 - B42$$
$$W123 - B135 - G246 - B142 - G365 - B263 - G154 - W456$$

In this way, eight more edge sets and two more vertex sets may be found. The Euler characteristic of the resulting 3-complex vanishes, because it has three cubical chambers and vertices, and nine faces and edges.

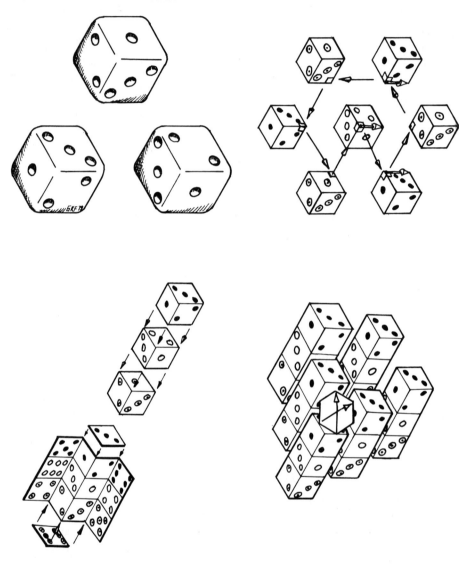

Figure 4   Dice

This 3-manifold inherits the Euclidean geometry of the cubes. Note how four chambers fit together at each edge without bending or leaving gaps. (Of course, two of the four chambers must be the same, i.e. the edge abuts one chamber twice.) Eight dice fit together neatly at the vertex. To discover the topology of this manifold, slide three dice together, 4(21). The surface has been peeled off to show the gluing pattern. This box is a fundamental region for the action on 3-space by the free abelian group Z^3 generated by three independent translations. In addition to the translation of $\sqrt{3}$ edge lengths normal to the picture plane, there are two independent translations by $\sqrt{2}$ parallel to the picture plane. These two generate the hexagonal pattern on which the illusion is based, 4(22). Thus the tribar lives in the geometrically flat 3-dimensional torus $T^3 = R^3/Z^3$, and there is a natural action of the cyclic group Z_3 of order three that "rotates" the tribar.

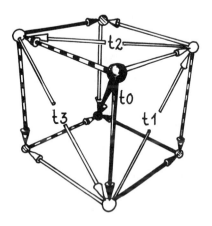
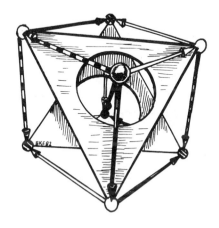

POINCARÉ GROUP

Figure 5

For a final look at the tribar, use the dice illusion to determine a gluing recipe on the surface of single cube. In effect, ignore the color labels B, W, G in the foregoing discussion. Identify faces 1-5, 2-4, 3-6. To see which edges belong together, I have colored them black, white and grey in 5(11). You may think of this skeleton cube as the white one in 4(12). The color labels on the vertices merely aid the eye, 5(12), because all vertices become one. Since the resulting manifold M has a single chamber and one vertex, Poincaré's edgepath method for computing its fundamental group applies. The fundamental group, $G = \pi_1(M)$, is generated by the three loops b, w, g, corresponding to the black, white and grey edges. The faces yield the relations

$$b^2 = wg \, , \; w^2 = gb \, , \; g^2 = bw \, . \tag{1}$$

The diagonals of the faces satisfy the relations

$$t_1 = bg^{-1} = b^{-1}w \, , \; t_2 = wb^{-1} = w^{-1}g \, , \; t_3 = gw^{-1} = g^{-1}b \, . \tag{2}$$

These could be regarded as defining three elements in the abstract group presented by three generators and relations (1). What follows is a geometric illustration of some basic combinatorial group theory. In the isometrically drawn cubical lattice any two of the t_i correspond to translations in the picture plane that generate the hexagonal tesselation, or more precisely, the underlying triangular lattice I used for the isometric drawings. The following relations can easily be checked algebraically:

$$1 = t_3t_2t_1 = t_1t_2t_3 = [t_1,t_2] = [t_2,t_3] = [t_3,t_1] \qquad (3)$$
$$t_2 = bt_1b^{-1} = wt_1w^{-1} = gt_1g^{-1}$$
$$t_3 = gt_2g^{-1} = bt_2b^{-1} = wt_2w^{-1}$$
$$t_1 = wt_3w^{-1} = gt_3g^{-1} = bt_3b^{-1}$$

The main diagonal of the cube corresponds to another element, t_0, in G, which commutes with everything because

$$t_0 = b^3 = w^3 = g^3 = wgb = gbw = bwg . \qquad (5)$$

If you remember your algebra you will note that, by (3) and (5), the t_i generate a free abelian group K of rank three. From (4), K is a normal subgroup of G. It follows from (2) and (5) that the quotient H of G by K is a cyclic group of order three represented by the cosets K, sK, s^2K where s is any one of b,w,g. Thus,

$$K \cong Z^3 , K \triangleleft G , G/K = H , H \cong Z_3 .$$

The group G illustrates the two fundamental theorems on "solving equations" over groups (in analogy to the classical theory of field extensions) discussed on Page 49 of Lyndon-Schupp [1977]. The first is a theorem by B.H.Neumann concerning the existence of roots (here the cube root of t_0) and the second is the starting point for Higman, Neumann and Neumann theory. It says that for any isomorphism between two subgroups of a given group there exist embeddings of the group in larger ones, called *HNN-extensions*, where isomorphic elements in the subgroups become conjugate under a soc-called *stable element* in the extension. You will find a good introduction to these matters in the textbook by Joe Rotman [1984].

Start with the free abelian group K as presented above. The permutation

$$t_3 \to t_2 \to t_1 \to t_3 , t_0 \to t_0 \qquad (6)$$

leaves the defining relations (3) invariant and therefore determines an automorphism of order 3 of the group K. Now adjoin a new element s to K, together with the relations

$$t_2 = st_1s^{-1}, t_3 = st_2s^{-1}, t_1 = st_3s^{-1}, s^3 = t_0 . \qquad (7)$$

You should check that this extension of K is isomorphic to the original presententation of G. The last relation in (7), together with (5) suggests that you send s to one of b,w,g. The three choices are related by the automorphism of G that permutes the generators $g \to w \to b \to g$. As you can check from (2) this automorphism of G restricted to K is the one given by (6).

CHAPTER 4 THE IMPOSSIBLE TRIBAR

Figure 6 CUBIC LATTICE

CUBIC LATTICE. *Figure 6.*

Geometers will by now have recognized G as a crystallographic group, and our manifold M as one of the six compact, orientable Euclidean space forms. These flat, Riemannian 3-manifolds are classified by their *linear holonomy group H*. You can learn about these matters in Wolf [1977]. In this context, the fundamental group G of M acts on its universal covering space, here Euclidean 3-space, as a group of Euclidean motions leaving the cubical lattice 6(11) invariant. Note how this perspective drawing dispels the mysterious illusion in the isometric view 6(12). Interpret the t_i as translations and s as the screw motion whose rotation angle is 120°, translation distance is $\frac{1}{3}\sqrt{3}$ edge lengths, and direction is the same as that of t_0. In 4(11) you may apply the screw at the barycenter of the three dice. In 6(11), one screw axis passes through the space between the grey, black and white dice.

Dividing out by the action of the translation subgroup yields the 3-chambered home of the tribar, the 3-dimensional torus with a Euclidean geometry. Its fundamental group is $K \cong Z^3$. The quotient of the torus by the action of the holonomy group $H \cong Z_3$, yields the Euclidean space form M, whose fundamental group is G, the extension of K by H.

MAN IN A CUBE. *Figure 7.*

The 3-dimensional world M that we have discovered, is just like ours in some ways, but very strange in others. The angles of plane triangles still add up to 180° and the familiar constructions of Euclidean geometry can still be carried out. But, it is a finite world, having the volume of a single cube. No matter how erratically you fly about it, there are points to whose vicinities you will have returned infinitely often by the "end of time." (The space is compact). As you look in certain directions you will see, at fixed intervals, first the top of your head, then your right side, then your back. Can you determine this direction on the cube? Thus, there is a straight line that closes on itself at right angles. Yet, this loop cannot be shrunk to a point. It gets stuck in the topology of the manifold. The tribar in the 3-fold, toroidal covering of M, however, has collapsed to a "unibar" with a single corner. Can you rebuild it in M with three corners?

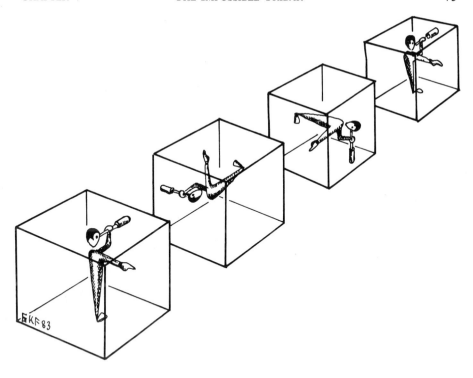

Figure 7 MAN IN A CUBE

Because we are 3-dimensional creatures, we find such spaces hard to imagine, and harder yet to investigate their properties. That is what topology is for. Two-dimensional worlds that are different from the Euclidean plane are easier, because we can make model surfaces of them. But even here, three dimensions are not always enough. That is what my next picture story is about.

Figure 8 LEFT HAND/RIGHT HAND

5
SHADOWS FROM HIGHER DIMENSIONS

A hemispherical bowl has a single, circular edge. So does a disc. If you sew the disc to the hemisphere you obtain a closed, two-sided surface; it has an inside and an outside. Topologists still call this a sphere, and it could be inflated back to a geometrical sphere if you like. Now take a long, narrow strip of paper, give it a half twist and glue the ends together. The resulting surface is not only one sided, it has but a single, closed edge. What happens if this edge is sewn to the rim of a disc?

Studying the variety of answers to this question, first asked by the astronomer August Möbius [1867], has been the gateway to topology for generations of students. Geometry is a "picture-friendly" branch of mathematics, or at least it was so a hundred years ago. Möbius' contemporaries found a model for his surface among Steiner's many brilliantly conceived but poorly "documented" geometrical constructions. Jakob Steiner, Pestalozzi student, tutor and "high school" teacher, colleague of Abel, Crelle and Jacobi, and ultimately holder of a special professorial chair in Berlin created by the brothers Humbolt, is the grandfather of *synthetic geometry*. This was a reaction to the *analytical geometry* that ruled the day. He left behind volumes of tantalizing theorems which were ultimately proved by Cremona and Veronese, using the very methods of analysis and algebraic geometry Steiner detested. Till the end of the century the best answer to Möbius' question was the surface Steiner called his *Roman surface* in memory of a particularly productive sojourn in that romantic city.

The only Hilbert student who wrote a dissertation in geometry, Werner Boy [1901], found a "simpler" surface in 1900. It is an immersion of the projective plane in 3-space, but Boy could not find the algebraic equation for his surface. Only recently, François Apery [1984] succeeded in doing that. Apery is a student of that great topological visualizer Bernard Morin of Strasbourg, to whom descriptive topology owes the greatest debt. Morin's vivid, pictorial description of his bold constructions has inspired their realization in many a drawing, model, computer graphic and film. But he insists that ultimately, pictorial descriptions should also be clothed in the analytical garb of traditional mathematics. What follows is an introduction to Morin's descriptive method.

This chapter is about curved surfaces in space; what they look like, how to draw them, and the algebra that is manifested in their construction. I also hope to teach you how to draw spatially extended shapes on paper and blackboard. The use of 2-dimensional *diagrams*, composed of clearly labelled and textually interpreted plane arrangements of points and lines, is well established in mathematics and taught in school. It is another matter with 3-dimensional *models*. To be sure, cylinders, globes and cones, the Platonic solids and an occasional Möbius band are all part of our students' mathematical furniture. These shapes can be reliably referred to by name, without further visual aid. This is not so for shapes the least bit more complicated. Here, the finished object, made with sticks and strings, or Plaster of Paris, or just its photographic image, is often all we show to the student. It is, however, in the making of the model, in the act of drawing a recognizable picture of it, or nowadays, of programming some interactive graphics on a microcomputer, that real spatial understanding comes about. It does this by showing how the model is generated by simpler, more familiar objects, for example, how curves generate surfaces.

CROSSING A CHANNEL. *Figure 1.*

This part of the story begins with a parabolic cylinder generated by a parabola inscribed in a rectangle. The parabola passes through two adjacent corners of the rectangle and is tangent to the opposite edge at its midpoint. Figure 1(11) embodies a recipe for sketching this curve inside a given perspective rectangle. Draw the diagonals of the quadrilateral (rear panel) to find its center. Connect the center to the two vanishing points (arrows). This halves the edges, thus quartering the rectangle. Diagonalize again (middle panel) to find the centers of the lower quarters or guess their location. These points lie on the parabola, and they determine the quarter points on the base edge of the rectangle. The lines from these quarter points to the corners are tangent to the parabola. The arcs, shown on the front panel, are now easy to draw. If necessary, the foregoing procedure may be iterated in the new, smaller rectangle determined by the quarter points. It juts out of the front panel for emphasis.

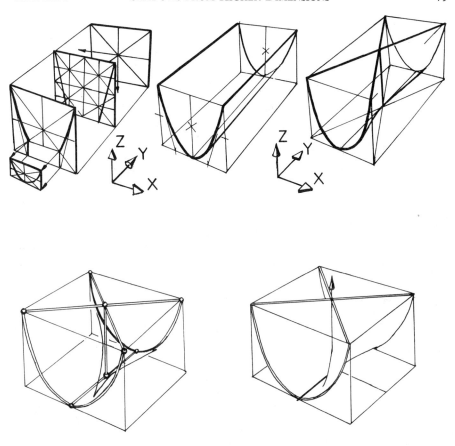

Figure 1 CROSSING AN CHANNEL

Now translate the *XZ*-rectangle, together with its parabola, orthogonally to itself, in the *Y*-direction. This forms the box which frames the surface. To draw 1(12) in a given box, construct parallel parabolas on opposite faces and connect corresponding points by straight lines. Note that the line traced out by the vertex of the parabola, call it the *keel*, is not in general the visible contour of the surface. I have drawn the *xy* and the *XYZ* coordinate frames separate from the origin of these coordinate systems. The origin lies on the keel, not, as is often drawn, on the contour. Note how the design of the arrowheads on the coordinate frames helps tell the story. The second *XYZ*-frame suggests that the surface is the graph of a function of two variables. Now make the surface cross through itself by reversing the parametrization of the front parabola, 1(23). Think of linearly connecting

$$
\begin{array}{lll}
X = -x & X = +x & X = xy \\
Y = -1 \quad to & Y = +1 \quad via & Y = y \\
Z = x^2 & Z = x^2 & Z = x^2.
\end{array}
$$

Column three is a canonical parametrization of the *Whitney umbrella* surface, whose topology was described in Figure 1:3. For sketching an umbrella, remember that it is attached to its box along the keel. The keel splits into two crossing lines that become diagonals on the face opposite that of the keel 1(21). The line of double points, call it the *stalk*, is orthogonal to the keel at the pinch point. In a sense, the stalk continues past the pinch point to form the "handle of the umbrella", also called its *whisker*. An algebraic geometer would eliminate the parameters, and consider the variety (= set of zeros) of $X^2 - Y^2Z = 0$ over the complex numbers. The negative Z-axis in our "real" space is where the "imaginary", and therefore invisible sheet of the surface crosses through itself. In Figure 2:11 you met a more complicated example of a whisker in connection with Morin's pinch point cancellation move. It plays an important role also in Apery's story, told at the end of this chapter. Let us return to the drawing lesson.

The keel joins the vertices of the two parabolas. You can find the precise location of the contour by connecting corresponding points on the parabolas by straight construction lines and sketching their envelope. A string model built inside a clear plastic box can be used to cast shadows illustrating this construction. A few rules of thumb can help you guess a plausible placement of the contour. Curves on the surface cannot cross a contour because the surface bends away from view there. Thus the contour is tangent to the parabolas and to the keel at the pinch point. Hence it is transverse to the stalk. You should practice drawing Whitney umbrellas for various positions of the box. Note that for 1(21), the contour is obliged to form a cusp, in which case the pinch point becomes invisible, 1(22).

You may have observed an *abus de dessin* in Figure 2:1(23), where I used the same straight line for keel and contour on the upper umbrella and skipped the contour entirely on the lower one. Such a graphical counterpart of Bourbaki's *abus de langage* is harmless provided the "errors" can easily be corrected and they do not mislead the viewer. Figure 2:1(23) is a drawing of two umbrellas joined crosswise to form a singular Möbius band which should, more properly, be called a *Möbius disc*. The surface is one-sided provided you consider the double line to be penetrable in some sense. My next figure explains how to do this.

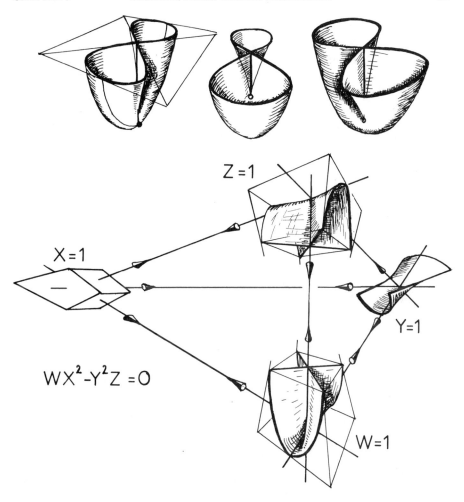

Plücker Conoid

Figure 2

For drawing pinch points on more general, flexible surfaces, it is important to distinguish two cases. First, when the umbrella bends towards its whisker, thereby crossing the plane tangent to the surface at the pinch point (the one transverse to the stalk), and second when it does not. To model this, bend the keel into the shape of a parabola in the stalk-keel plane by changing $Z = x^2$ to $Z = x^2 \pm y^2$. After eliminating the parameters to obtain $X^2 - Y^2Z \pm Y^4 = 0$ you can study the cross-sections of the new shape by holding Z constant. For a negative sign the result looks like the shadow of a branch point, as in Figure 1:5(12). Note how the contour develops two cusps near the pinch point. The contour would close on a third cusp further up the stalk if the surface were to open wider, and you could see the stalk from inside the channel.

When the sign is positive and the umbrella is entirely on one side the tangent plane at the pinch point, the constant Z sections, $X^2 = Y^2(Z - Y^2)$ are figure-eights. Some conventional views are in 2(11) to 2(13). Actually, drawing the contour cusps on the stalk of 2(11) and 2(12), instead of very near the stalk tangency, is another *abus de dessin*. Detail 2(13) has a more serious flaw, in that it violates the rule that pinch points belong on contours. Can you discover the topology of this surface? It is not a disc.

Let us return to the straight umbrella and "homogenize" the cubic polynomial $X^2 - Y^2Z$ we obtained earlier thus: $WX^2 - Y^2Z = 0$. This is the equation of a *Plücker conoid*. The is surface in projective 3-space with homogeneous coordinates $W:X:Y:Z$. If you interpret Cartesian XYZ-space as the finite part obtained by setting $W = 1$ you have what we started with.

Recall that a projective space of two, three or more dimensions may be defined as the totality of all straight lines through the origin of a Euclidean space of one dimension higher. To visualize what the projective umbrella looks like at the "end" of the stalk and of the keel, take a different "picture" of projective space. This time capture the lines through the origin in 4-space on a 3-space cross section given by $W + X + Y + Z = 1$. The coordinate axes of 4-space cross this "hyperplane" at the vertices of a tetrahedron, and we can interpret W,X,Y,Z as barycentric coordinates in that 3-space. The tetrahedron is drawn in the lower half of the figure, and the surface details show what it looks like near the "four corners of the world". This picture is a *perspective chimera* in that it has several different triplets of vanishing points. The special effect that can be achieved with multiple perspective is familiar to fans of Escher's drawings. Here I have drawn the umbrella at the lower corner of the tetrahedron as if $W = 1$ held in an entire cubical neighborhood. So I used the three other corners as the vanishing points of this detail; and similarly at the other corners.

Each line through the origin of a Euclidean space also pierces the unit hypersphere in two antipodal points. The sphere is thus a double-cover of the projective space. A hemisphere covers the projective space once, except along its equatorial border. If this border were smoothly sewn together so as to identify antipodes, a topological model of the projective space would be realized. In dimension two, such a surface was proposed by Möbius [1867] as an example of a closed, one sided surface. He imagined it as a closed ribbon with an odd number of half twists, sewn to a disc along its boundary. Of course, no such tailoring is possible in 3-space: we need an extra dimension.

CHAPTER 5 SHADOWS FROM HIGHER DIMENSIONS 83

Consider first a parametrization

$$
\begin{aligned}
W &= z^2 \\
X &= xy \\
Y &= yz \\
Z &= x^2
\end{aligned}
$$

of the conoid obtained by "homogenizing" the formulas for the umbrella. Suppose the source of this parametrication is the unit sphere and that two points (x,y,z) and $(\bar{x},\bar{y},\bar{z})$ on the sphere are mapped to the same point in 4-space. To see that the two points must be antipodes on the sphere let me list nine relations.

$$
\begin{array}{lll}
x^2 = \bar{x}^2 & yz = \bar{y}\bar{z} & \bar{x} = \pm x \\
y^2 = \bar{y}^2 & zx = \bar{z}\bar{x} & \bar{y} = \pm y \\
z^2 = \bar{z}^2 & xy = \bar{x}\bar{y} & \bar{z} = \pm z \ .
\end{array}
$$

From $x^2 + y^2 + z^2 = 1$ on the sphere it follows that any two relations in the first column implies the third. Each of these implies the corresponding relation in the third column. As yet we cannot claim that all three signs are matched. Complex numbers arithmetic leads to $(x + iz) = \pm (\bar{x} + i\bar{z})$ from $yz = \bar{y}\bar{z}$ via $(x + iz)^2 = (\bar{x} + i\bar{y})^2$. A similiar argument shows that the first and third signs in column three also match, hence they match altogether. (Note, for the sake of symmetry, that the third relation in the second column now also follows.)

Thus we have parametrized a projective plane embedded in Euclidean 4-space and shown that the surface described by Möbius can be found there. It is a shadow of the *Veronese surface*, which is, in turn, an embedding of the projective plane in projective 5-space given by

$$ x : y : z \rightarrow x^2 : y^2 : z^2 : yz : zx : xy \ . $$

PINCHING A ROMAN SURFACE. *Figure 3.*

Here is another, more beautiful projection of Veronese's surface into 4-space. For notational convenience, I shall double the size, and give its parametrization in Cartesian, cylindrical and spherical coordinates, using the abbreviations S,C for sines and cosines. The equatorial angle is θ and α is the angle from the equator towards the north pole on the unit sphere.

$$
\begin{aligned}
W &= x^2 - y^2 = r^2 C(2\theta) = C(\alpha)^2 C(2\theta) \\
X &= 2xz = 2rC(\theta)z = S(2\alpha)\, C(\theta) \\
Y &= 2yz = 2rS(\theta)z = S(2\alpha)\, S(\theta) \\
Z &= 2xy = r^2 S(2\theta) = C(\alpha)^2 S(2\theta) \ .
\end{aligned}
$$

84   A TOPOLOGICAL PICTUREBOOK

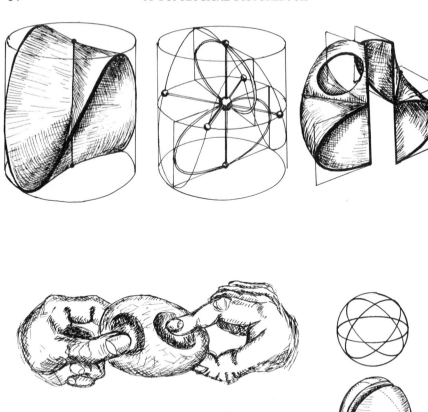

Figure 3   PINCHING A ROMAN SURFACE

To see that this is also an embedded projective plane, proceed as before to show that two points on the unit sphere mapped to the same point in 4-space are antipodes. From the second column observe that

$$(x + iy)^2 = (\bar{x} + i\bar{y})^2$$
$$(\bar{x} + i\bar{y}) = \pm (x + iy)$$
$$(x + iy)z = (\bar{x} + i\bar{y})\bar{z}$$
$$(\bar{x}, \bar{y}, \bar{z}) = \pm (x, y, z).$$

Let us first consider the *XYZ*-shadow of this surface and begin with the unit θ-circle in the horizontal *XY*-plane. Hold $r = 1$, $z = \frac{1}{2}$ constant and note how $Z = S(2\theta)$ bobs up and down twice, which bends the circle into a *wobbly hoop* as in Figure 1:2(21). Diametrically opposite points on the circle reach the same altitude on the hoop, and these four points determine a vertical rectangle, one for each pair $\pm \theta$. As the rectangle turns inside the vertical cylinder, its height changes sinusoidally, 3(12). Twice it degenerates into a diameter of the unit circle.

The mobile edge of this rectangle, connecting opposite points on the wobbly hoop, generates a ruled surface, 3(11), consisting of two Whitney umbrellas joined as in Figure 2:1(23). This Möbius disc is a special projection of Plücker's conoid; a calculation I leave to you. Use the cylindrical parametrization, third column, to generate this surface, holding $r = 1$ constant and releasing z to go from $+ \frac{1}{2}$ to $- \frac{1}{2}$. If, instead, you release r and hold z constant, you inscribe a flexing parabola in the rectangle, which generates the saddle surface, Figure 1:2(22). Together, these two surfaces form a topological model of the *Roman surface*. The algebraic surface is obtained by rounding the rotating rectangles into ellipses, 3(12). This bends the two umbrellas and flattens the parabolic saddle.

To verify this, decompose the spherical parametrization thus

$$\begin{vmatrix} C(\theta) \\ S(\theta) \\ 0 \end{vmatrix} S(2\alpha) + C(\alpha)^2 \begin{vmatrix} 0 \\ 0 \\ S(2\theta) \end{vmatrix}.$$

The sliding coefficients of this interpolation between two vectors, $A = S(2\alpha)$ and $B = C(\alpha)^2$, satisfy the equation of a semi-ellipse inscribed in a unit square, $A^2 = B(1 - B)$. To see this geometrically, note that the rescaled equation $(\frac{1}{2}A)^2 = (1 - B)(B - 0)$, expresses *Thales' principle*: as a point B moves across the diameter of a semi-circle, the altitude over B is the geometric mean of the distance from B to the ends of the base. Since circles go to ellipses under an affine projection, such an interpolation between any two vectors produces the semi-ellipse inscribed in the associated parallelogram. This interpolation thus joins the more familiar trigonometric and linear interpolations, based on $A^2 + B^2 = 1$ and $A + B = 1$, which are so useful in spherical and projective coordinatizations.

The process also creates four new umbrellas, 3(13), whose stalks are the two degenerate ellipses. The three double lines cross at a triple point and end at six pinch points. Here are some ways to model the Roman surface. In the first, pinch dimples on opposite sides of a clay ball, 3(21), and imagine the place for the six pinch points. Progressively mold the pinch points in place as you rotate the model a third of a turn at a time. A piece-wise linear model, called the *heptahedron*, may be built out of three mutually orthogonal squares which cross each other on diagonals. Imagine Cartesian coordinate planes inside an octahedron, with opposite octants capped by triangles, or look at Figure 288 of Hilbert and Cohn-Vossen [1932]. A quick way to produce something "half way" between the heptahedron and the smooth surface from a ball of art gum eraser is shown on the right, 3(33). Take two quarter sections of the ball and cut them almost in half. Now turn the eighths against each other and stick the two pieces together. It looks like a ball with alternate octants removed.

GAUSS MAP AND CROSS CAP. *Figure 4.*

When modelling or drawing the Roman surface, remember that it fits inside a tetrahedron, touching the faces along circles. The curvature of the surface changes at these four circles 3(31) and 3(32). Informally speaking, a surface bends to one side of its tangent plane in a neighborhood of a point of *positive curvature* and one says that the surface is *supported* by the plane. The surface crosses its tangent plane at a point of *negative curvature*.

To verify the first assertion algebraically, consider where the slant planes of the tetrahedron meets the surface. For example,

$$1 + Z - X - Y = x^2 + y^2 + z^2 + 2xy - 2xz - 2yz = (x + y - z)^2 \geq 0 .$$

Since this is non-negative, the slant plane supports the surface on a curve which is the image of the circle on the source sphere cut by the plane $z = x + y$. To discover the shape of this curve consider the calculation

$$1 = (X + Y - Z)^2 = X^2 + Y^2 + Z^2 + 8xyz(z - x - y)^2 = X^2 + Y^2 + Z^2 .$$

It shows that the distance from the origin in XYZ-space to this plane curve is constant; so it is a circle.

The curvature changes sign along what are called the *parabolic curves* on a surface. There is a practical way of locating certain paraboloic curves on a surface, call it the *Hessian principle*. I shall use it to establish the second assertion. Let $F:R^2 \to R^3$ parametrize a surface in space. The cross-product $N = F_x \times F_y$ of the two tangent vectors gives a vector normal to the surface. A change of coordinates multiplies N by the *Jacobian determinant*:

$$F_u \times F_v = (F_x \times F_y) \frac{\partial(x,y)}{\partial(u,v)} .$$

Chapter 5 — Shadows from Higher Dimensions

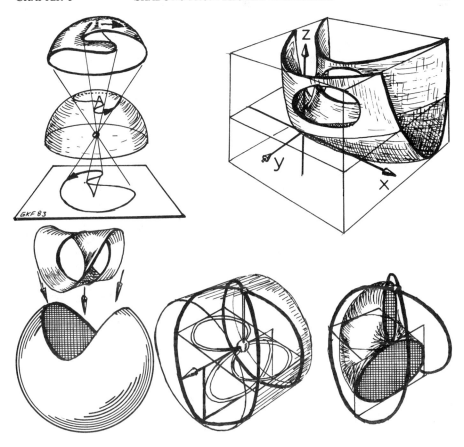

Figure 4 GAUSS MAP AND CROSS CAP

Thus the *unit normal* $N/|N|$ is independent of the parametrization and defines a mapping from the surface to the unit sphere, called the *Gauss map*. The Jacobian of the Gauss map is called the *Gauss curvature function* on the surface. Here is a less complicated way of thinking about this function. For each plane normal to a surface at a point, the curvature of the curve of intersection is a *sectional curvature*. The Gauss curvature is the product of the maximum and the minimum sectional curvatures at the point. (The average sectional curvature is the *mean curvature* and this vanishes on surfaces minimizing area locally; soap films, for example.)

For drawing or modelling it usually suffices to know where the curvature is positive or negative and to find the parabolic curves on the surface. Suppose, for the moment, that the surface is given locally as the graph of a function $z = f(x,y)$. The surface bends to one side of its tangent plane where

$f_{xx}f_{yy} - (f_{xy})^2$, the *Hessian determinant* of $f(x,y)$, is positive. The surface crosses the tangent plane where the Hessian is negative. Though you can derive this from the second Taylor expansion, I prefer to argue geometrically as follows. The vector $N = \langle -f_x, -f_y, +1 \rangle$ is normal to the surface and the Gauss map, $(x,y) \to N/|N|$, takes its values in the upper unit hemisphere. Since central projection preserves orientation, see 4(11), you need only check the orientation of the plane map $(x,y) \to (f_x, f_y)$. The Jacobian of an orientation preserving (reversing) mapping is positive (negative). Thus, the Gauss map and the Hessian have the same sign.

Let us apply this principle to a general model of a surface patch obtained by bending the keel of a U-shaped channel, $z = u(x)$, $u'' > 0$, horizontally to $y = a(x)$, $a'' > 0$, and perhaps also vertically, $z = b(x)$, see 4(12). (All variables have zero derivative at the origin.) Compute:

$$z = f(x,y)$$
$$f = u(y - a(x)) + b(x) \qquad \operatorname{grad}(f) = \begin{vmatrix} f_x \\ f_y \end{vmatrix} = \begin{vmatrix} b' - a'u' \\ u' \end{vmatrix}$$

$$\begin{vmatrix} f_{xx} & f_{xy} \\ f_{yx} & f_{yy} \end{vmatrix} = \begin{vmatrix} b'' - a''u' + (a')^2 u'' & -a'u'' \\ -a'u'' & u'' \end{vmatrix} = \begin{vmatrix} b'' - a''u' & 0 \\ 0 & u'' \end{vmatrix}$$

In the case of the Roman surface, the keel remains in a plane; so $b'' = 0$ in the model. Thus the curvature changes exactly on the keel $y = a(x)$. In general, the parabolic curve shifts to one side of the keel according to the sign of $b''(x)$. On the surface 4(12) the parabolic curve shifts to the outside because $b'' < 0$. For example, the *Peano saddle*

$$f(x,y) = (y - x^2)(y - 3x^2) = (y - 2x^2)^2 - x^4,$$

changes curvature on the parabola $2y = x^2$, on which $\operatorname{grad}(f) = (8x^3, -3x^2)$. In other words, the Gauss map develops a Whitney cusp, as discussed in Chapter 1. You will find a very informative treatment of these matters in the monograph on the cusps of Gauss maps by Banchoff-Gaffney-McCrory [1982].

The *YZW*-shadow of the Veronese surface from 4- to 3-space is another important surface of Steiner, the *cross cap surface*. This time, the θ-circle, $(S(\theta), S(2\theta), C(2\theta))$, wraps twice about the *Y*-axis, forming a *double hoop*, 4(23). The double hoop is usually associated with the complex squaring function. You can use the Cartesian parametrization of the Veronese surface I gave above to visualize this Riemann surface in *XYWZ*-space by holding $z = \frac{1}{2}$ constant and letting $x + iy$ range over the complex plane.

As before, connect each $\pm \theta$ places on the hoop by a line segment parallel to the *Y*-axis and complete the rectangle. This time the rectangle rotates about the one of its sides, and degenerates once as it squeezes through the positive *W*-axis. Now release the α-parameter, which, under Thales' principle, inscribes an ellipse in the rectangle, 4(23). The completed surface looks like a Möbius disc glued into a spherical shell, 4(21).

The cross cap is supported by the dihedral $W \pm \sqrt{2}Y = 1$, which fits over it like a roof. The curvature changes at the two ellipses where the cross cap touches the roof, by the Hessian principle. You will find a more complete description of the cross cap, with many illustrations, in Chapter VI of Hilbert and Cohn-Vossen [1932]. The computer animation by Banchoff and Strauss [1977] shows the transition from the Roman surface to the cross cap, effected by a rotation in 4-space. This film dramatically reveals the many remarkable shadows this beautiful surface casts into our space.

BOY SURFACE. *Figure 5.*

During the latter half of the 19th century the theory of surfaces matured and flowered. Its influence on the topology of manifolds is nicely documented in Scholz's history book [1980]. At the turn of the century, however, there remained the tantalizing question of what a projective plane might look like if it were immersed (no pinch points) in 3-space. A topological solution was found by Hilbert's student Werner Boy [1901]; but he was unable to satisfy the conventional expectations of his day: to give analytic formulas for his surface. Only in the last decade have Bernard Morin and his collaborators succeeded in that task. But that is getting ahead of our story. You will find photographs of a wire mesh model of Boy's surface in Hilbert and Cohn-Vossen [1932, Fig. 321a-d], along with an introduction to the topology of this surface, which I shall not repeat here.

It is natural to ask whether, like Steiner's two surfaces, Boy's surface is also the 3-dimensional shadow of a projective plane embedded in 4-space. The negative answer has an elegant demonstration, which I learned from Ben Halpern and Jeff Boyer. The preimage, C, of the double curve, K, of an immersion $F:RP^2 \to R^3$, is connected because K has a triple point. Let $f:S^1 \to RP^2 ; f(S^1) = C$ denote a parametrization of this immersed circle for which $F(f(\theta)) = F(f(\theta + \pi))$. The difference function $w(\theta) = W(f(\theta + \pi)) - W(f(\theta))$, where W is the fourth coordinate of a presumed embedding of RP^2 in R^4, is never zero. But if $w(\theta) > 0$ somewhere, then $w(\theta + \pi) = -w(\theta) < 0$. By continuity it must vanish in between.

Since no suitable embedding exists, let us relax that condition, and see just how cleanly the projective plane can be placed in 4-space so that its shadow is Boy's surface. For observing 4-dimensional fauna in the absence of convenient analytic formulas we need some descriptive tools. The 2-dimensional shadow-watchers in Plato's cave could have benefited from perspective drawing to encode the 3-dimensional world outside. Let us place the colors of the rainbow at the disposal of 3-dimensional modellers so that they may *paint* the fourth coordinate onto their surfaces. Regrettably, book pages and blackboards are flat and most pictures drawn on them are monochromatic. So you will have to be content with windows and painting instructions.

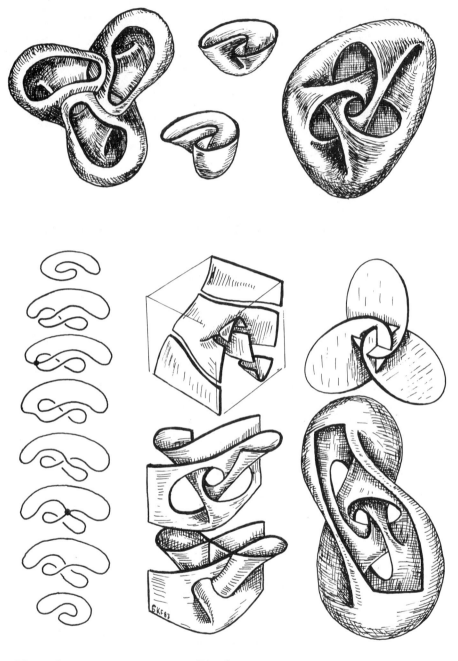

Figure 5 BOY SURFACE

To reduce the complexity of the line pattern, deform the conventional shape of Boy's surface 5(11) by *rolling* the cusp forward, as in detail 5(12). This splits the contour of 5(11), with its three cusps and three apparent crossings, into two curves, a *deltoid* inside an oval, as in 6:6(14). The double curve, a bouquet of three loops tied to the triple point, remains the same, once the windows in 5(13) are replaced. I have tilted the deltoid relative to the vertical direction so that the horizontal saddle is near one cusp. By rotating 5(11) slightly to cancel an extra saddle against a maximum, this surface also supports a Morse function with only three extrema.

Paint 5(13) green for ground level and redden the top window to eliminate the upper loop of the double curve in 4-space. Push both lower windows into the blue direction to clear the green surface. Since the rims of these two discs link in 3-space, the preimage of their intersection (the vertical slit in 5(33)) is an arc from border to interior in each disc. As the color of the arc in the right disc rises from the violet interior to blue near the border, the arc in the left disc falls from blue to violet. By continuity there is one point inbetween where both arcs have the same shade of indigo. This is where the *node* of the surface in 4-space occurs. Note that if, in 3-fold rotational symmetry, the top window had also been pushed into the blue direction, then two more indigo nodes would have been formed, one for each of Morin's pinch point cancellations on the Roman surface raised to 4-space.

Another way to build up dimensions is by stacking consecutive 3-D sections. Slice 5(13) by horizontal planes and reinterpret each plane as the horizontal of a 3-dimensional slice of 4-space. Eight sections are stacked in the left column of 5(21) to 5(91). I have used the convention of knot diagrams, interrupting a line as it passes under another. Note that "red" is "down" in each of these eight copies of 3-space. (I have omitted drawing the obvious point followed by an oval at the top, and the oval shrinking to a point at the bottom.) If you interpret the slicing plane of 5(13) as falling in time, then the stack depicts eight instances, $T_1...T_8$, of a curve moving in 3-space. It is a simple closed curve except at two instances. At T_3 a *recombination*, as in molecular genetics, signals a saddle orthogonal to the *(time)×(color)*-plane in 4-space. In my drawing, it has the shape of cubic surface, showing a cusp in the picture plane, 5(32). Later, at T_6, the curve crosses over itself, signifying the node on the surface in 4-space.

The shadow of this moving curve in the picture plane (close the gaps) depicts a regular homotopy of an immersed circle except at the recombination point. Stacking these temporal slices spatially, rebuilds Boy's surface from the top down. This is, in fact the technique orignally used by Boy. By T_1, a cusp has occurred in the contour. A curve of double points is created by T_2. Past the saddle, the curve has two opposing loops, T_4. A second double curve and a triple point is created as a segment sweeps across the left loop between T_4 and T_5. The last loop of the double curve disappears between

T_7 and T_8, and a cusp allows the surface to close up. The stack from T_4 to T_8 is the *crossed cusp* shown with windows, 5(42), and without, 5(52), which form the triple point, 5(23) of Boy's surface, as in 5(33) from Francis-Morin [1979].

SLICE AND SHADOW. *Figure 6.*

Let me review this technique of visualizing painted surfaces in 4-space by applying it to the 3-dimensional surface primitives, the pinch point, 6(22), and the double line, 6(42). Consider 4-space framed by an orthogonal triple, $R^2 \times R^1 \times R^1$, consisting of the *base plane* and the *temporal* and the *chromatic directions*. In practice, the base plane is usually also the picture plane of the illustrations. For clarity, I have tilted the base plane. It is represented by the front face of the boxes framing the details.

If you slice the surface in 4-space by moving the chromatic cube parallel to itself in the temporal direction, you will see a curve moving in space. For example, the motion 6(21)-6(12)-6(23) untwists an apparent loop, while 6(41)-6(32)-6(43) exchanges over and under crossing of the loop, signaling a *node*, which is a one-point self intersection of the surface in 4-space. Projected to the base plane, the former motion looks like a loop being pulled taut, passing through a cusp singularity. The latter projection does not move.

Stacking the base curves produces a Whitney umbrella, 6(22), and a crossed tube, 6(42), respectively. The pinch point in the umbrella corresponds to the cusp on the base plane shadow. On the other hand, a bit of calculus will convince you that the apparent cusp in the base projection of the loop indicates a tangency of the surface to the chromatic direction in 4-space. The node of the second surface is forced by the linking of the windows used to remove the double locus.

WHITNEY BOTTLE. *Figure 7.*

The fact that the motion of the slicing curve for Boy's surface, 5(21) through 5(91), casts no cusped shadow in the base plane, says that the unit red vector is never tangent to the surface in 4-space. In other words, this 4-dimensional immersion of the projective plane supports a continuous field of normal vectors, something no immersion of a closed non-orientable (1-sided) surface in 3-space can do. The embedding of such surfaces in 4-space which admit a furry pelt without bald spots was investigated by Whitney [1941] in the course of developing the theory of characteristic classes. He proved that for the case of odd Euler characteristic (such as P^2) no such embeddings exist. Thus the self-intersection cannot be removed by a smooth deformation through immersions (a regular homotopy) in 4-space.

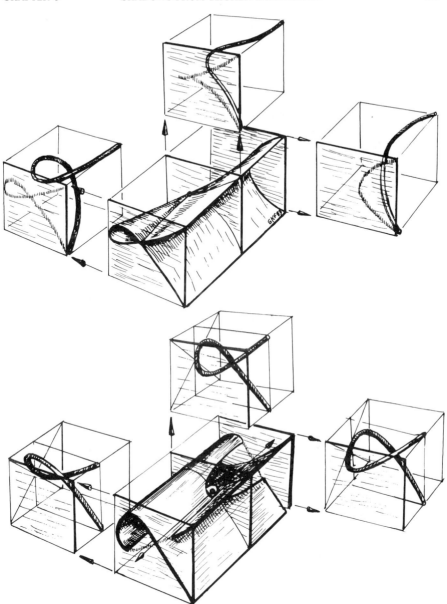

Figure 6 SLICE AND SHADOW

Figure 7 WHITNEY BOTTLE

Whitney's conjecture for the case of even Euler characteristic, proved by Massey [1969], implies, in particular, that the Klein bottle has three inequivalent embeddings. One of these is furry without bald spots. You can construct the Whitney stack for this one by slicing the usual immersion of the Klein bottle in 3-space. To visualize the Klein bottle without normal field, described in Figure 4 of Whitney [1941], proceed as follows.

First note that if at time T_4 in Boy's stack, 5(51), we had simply untwisted the two little loops, an embedding of the projective plane would have resulted. The temporal shadow of this projective plane would have a segment of double points ending in two pinch points. In other words, this is an isotope of Steiner's cross cap. You should design the Whitney stack, whose shadow is the Roman surface, as an exercise. The temporal section of Whitney's bottle, reading 7(11) to 7(91) from the top down, develops two loops that recombine across a saddle forming two linked ovals. Flip one of them over and reverse the process. A schematic assembly of details in the shadow of this surface, second column, may be read from the Whitney stack. Note the singular Möbius band crossed by a plane with a triple point corresponds to the flip of the oval. A complete assembly of the shadow is shown at the right, 7(13) to 7(53) using both windows and zones.

ROMBOY DEFORMATION. *Figure 8.*

There is a topological transition from Steiner's Roman surface to Boy's surface based on the the cancellation of adjacent pinch points, see Figures 2:9, 2:10 and 2:11. This deformation may well have been known to Boy and used by him to discover the immersion of the projective plane that bears his name. I learned it from Tim Poston in 1977, who learned it from Bernard Morin. For convenient comparison I have redrawn several views of the surfaces you have already met, together with their line patterns.

Begin with the conventional view of the Roman surface 8(12). In its line pattern 8(11) are two contours (solid lines), three double curves (dashed lines), one triple point (center), six pinch points and arrows indicating where to cut the surface. The section 8(22) contains the southern and south-eastern pinch points, which are cancelled first, 8(32). Globally, we have reached the line pattern 8(41) from which you can draw the surface 8(42). The western pinch points cancel next, 8(51) and 8(52). You can also imagine this surface as the result of untwisting two little loops, one after the other, on the second curve from the top in the stacked cross-sections of Boy's surface, 5(31). Once the last two pinch points cancel, 8(62), you are looking at the backside of 5(11), but in a mirror. Thus 8(63) is a mirror image of 8(62), and with three windows, 8(64), it is the same view of Boy's surface as 5(11), but rotated 45°.

The third column depicts the analogous sequence taking the "flat" view of Boy's surface 5(13) to the conventional "round" view. The northern cusp rolls over the contour, 8(23) to 8(33). This pair of cusps is an upside down view of the pair in 5(12). Can you tell, by looking at 8(44) and 8(54) which cusp has to roll over to obtain view 8(53)? Cusp rolling is an isotopy in 3-space, but cancelling pinch points is not necessarily the shadow of an isotopy in 4-space. As we have seen, in contrast to the Roman surface Boy's surface cannot be the shadow of an embedded projective plane in 4-space.

The two columns of this tableau also portray qualitatively the two methods used by François Apery to obtain explicit parametrizations of Boy's surface [1984]. For the first, recall that an ellipse generates Steiner's Roman surface. The sculptor Max Sauze built such a model for Bernard Morin. He decided to use planar ovals also to construct a wire model of Boy's surface. This inspired the physicist and mathematical artist, Jean-Pierre Petit and his collaborator, Souriau [1982], to program the surface on an Apple computer. Their ellipse changes its shape and plane but stays tangent to a base plane at the *south pole* of the surface. For the surfaces in the tableau the south pole lies on the window of 8(62) and is furthest from you in 8(63). It is your nearest point on 8(12) and is visible through the three windows of 8(13). Petit and Souriau empirically determined formulas for the the position of the ellipse in its plane and the tilt of the plane for each moment of the motion.

Apery begins, instead, with an elliptical generation of the Roman surface. His is different from the one I used above. In Apery's the ovals remain tangent to a supporting plane at the south pole. Consider a point moving on a horizontal circle of radius r_1 at height 1, going twice as fast and in the opposite direction as a point moving about a parallel circle of radius r_0 in the base plane. The south pole is the origin.

$$\begin{vmatrix} r_1 C(2\theta) \\ r_1 S(2\theta) \\ 1 \end{vmatrix} \frac{1}{1+t^2} + \frac{t}{1+t^2} \begin{vmatrix} + r_2 C(\theta) \\ - r_2 S(\theta) \\ 0 \end{vmatrix}$$

Apery takes the southern circle of radius $2/3$ and the northern circle at $1/3\sqrt{2}$. For each θ consider the vector $J(\theta)$ from the origin to the upper point and the horizontal vector $K(\theta)$ in the base plane to the lower point. The sliding coefficients $A = 1/(1 + t^2)$ and $B = tA$ satisfy Thales' principle. Therefore the ovals generated thus: $J(\theta)A(t) + B(t)K(\theta)$, are ellipses tangent to the horizontal plane at the origin. The three components of this parametrization are quadratic polynomials in the Cartesian coordinates of the unit sphere. The parameter t is the tangent of the latitude on the sphere.

Apery ingeniously perturbs the sliding coefficients A,B by deforming their common denominator to $1-2\beta t + t^2$ where $\beta = (b/\sqrt{2})\sin(3\theta)$. This has the remarkable effect of cancelling all six pinch points at $b = 1/\sqrt{3}$. In an algebraic

Chapter 5 — Shadows from Higher Dimensions

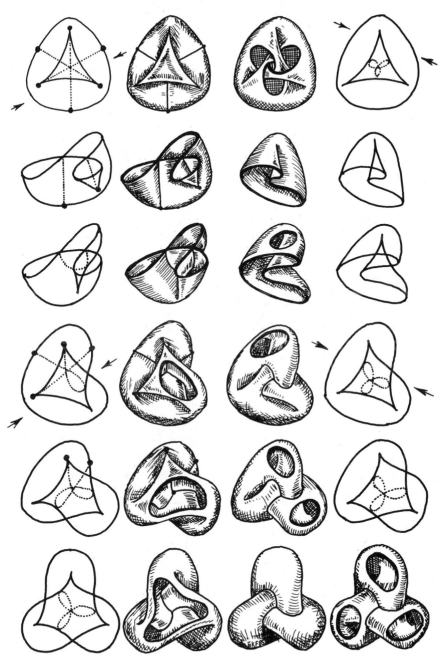

Figure 8   Romboy Deformation

tour de force, he eliminates the parameters and obtains a polynomial equation of degree six for Boy's surface.

For reasons derived in classical algebraic geometry, Boy's surface must have degree at least six. In a review of Boy's problem, Heinz Hopf [1955, 1983 p.104] conjectured that a polynomial parametrization would also be at least of degree six. Apery, however, found a parametrization of Boy's surface using homogeneous polynomials of degree four. Here, his approach is more closely related to Morin's original way of finding parametrizations of sphere eversions [1978]. The visualization, both graphically and analytically, of these complicated surface homotopies is the subject of the next chapter. I shall conclude here with a thumbnail sketch of Apery's second surface.

Morin's method is the analytical analogue to the graphical technique I described in Chapter 2. Starting from the contour, build up the surface from this 2-dimensional template. Apery begins by stabilizing a singular map of the sphere into the plane so that the image of the critical points looks like 8(14), without the dotted curve. You may have noticed that the contour image of the cusp-rolling motion resembles the sequence of pictures associated with Thom's *hyperbolic umbilic catastrophe*. More about this resemblance, which is not coincidental, is amply illustrated and patiently explained by Jim Callahan [1974,1977].

Once the cusp-rolling procedure is captured algebraically and given a three-fold symmetry, Apery has a motion from 8(14) to 8(61). The last step is to "inflate" this one parameter family of sphere-to-plane mappings. By chosing the third coordinates carefully so that the map from the projective plane into 3-space is an immersion at all stages he obtains an isotopy from 8(13) to 8(63).

The pleasant fact that all these surfaces and their deformations can now be coded in simple languages like BASIC, and viewed on inexpensive, easy to use graphics computers, like the Apple, places this medium on the same elementary level as the hand graphics in this picture book. That is why I have emphasized the analytic counterparts of some of the drawing techniques. Nonetheless, the limited resolution, slow speed and modest flexibility of computer graphics at a less than heroic hardware level benefits greatly from the programmer's ability to sketch effectively those views of the objects the computer is to render.

6
SPHERE EVERSIONS

The story of everting spheres in 3-space by regular homotopies is the case history of a nontrivial visualization problem of remarkable complexity and compelling beauty. The task is to show the motion of a spherical surface through itself in space so that, without tearing or creasing, the surface is turned inside out. That so many different graphical methods were applied to the same problem, when it is more in the nature of mathematics to display the versatility of one method by applying it to a variety of different problems, makes it a paradigm for descriptive topology. It is premature to formulate explicit ground rules for comparing the many solutions to this graphical problem. Instead let me tell a very short story about golden rectangles instead. Although the existence of golden rectangles has no direct bearing on eversions, the universal currency of this topic among mathematicians makes for a good example of the "graphical criticism" I have in mind.

GOLDEN RECTANGLE. *Figure 1.*

Recall that the sides of a golden rectangle are incommensurable because Euclid's algorithm does not terminate with a common measure for the sides. The geometrical form of the recursive step in the algorithm is to delete the square on the (shorter) side of a rectangle, and repeat the procedure on the remainder, if there is one. Now suppose there were an initial rectangle whose remainder, after separation from the square, is a smaller copy of itself. Then this *anthyphaereisis*, this reduction of the larger by the smaller, would never stop. But why is there such a rectangle in the first place? Here are three ways of answering this question. The first specializes an abstract theorem whose proof has, in fact, little to do with golden rectangles.

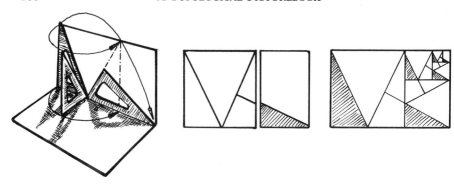

Figure 1   GOLDEN RECTANGLE

Let x denote the ratio of the longer to the shorter side of a rectangle. The defining proportion $x : 1 = 1 : x - 1$ leads to an equation $f(x) = 0$ where $f(x) = x^2 - x - 1$. By the calculus, this polynomial function increases from $f(1) = -1$ to $f(2) = +2$ because its derivative is positive on the interval. Differentiable functions are continuous, and the latter have zeros on intervals on which they change sign. But we don't really need the calculus, because by a well remembered formula from highschool algebra, $x = \frac{1}{2}(1 + \sqrt{5})$, is the root of the quadratic equation. But we don't need algebra either. Here is a picture proof that even grade schoolers can appreciate. Stand a right triangle, of side ratio 2:1, on its short leg as in figure 1(11). Tip one copy over in its plane to rest on its hypotenuse. Flip a second copy 180° out of the plane. The tipping maneuver determines a rectangle, of the correct ratio, by Pythagoras! The flipping maneuver marks the square to be taken away, leaving a smaller rectangle with a right triangle in place. Now convince your audience, by a demonstration tailored to their measure, that the procedure, applied to the smaller triangle, 1(12), fits perfectly inside the smaller rectangle, and therefore continues to do so forever, 1(13).

Thus, there is a kind of linear order imposed on a set of competing expressions of the same basic idea. In one direction there is a progression of ever more rudimentary picture fragments associated with an increasingly abstract and general theory. Usually, someone illustrating the demonstration based on continuity would scribble a wiggly graph from one corner of a rectangle to the opposite corner and call attention to how it must cross a horizontal line representing the x-axis. In the other direction there is a progression of ever more concrete examples with particularly memorable constructions and rich associations. Ideally, the route back to the general theory (by way of abstraction) should be illuminated by the brilliance of the special example.

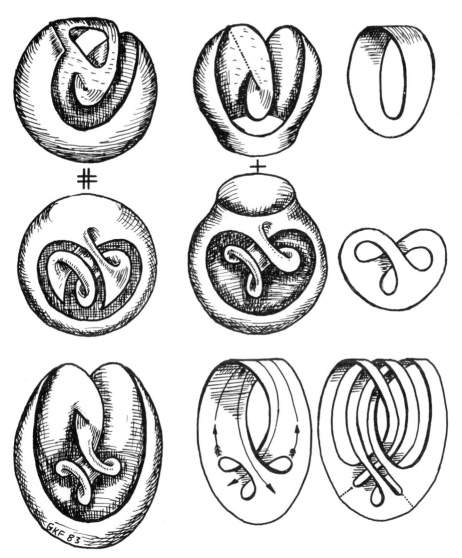

CROSS CAP AND HANDLE

Figure 2

A bit of history helps to point out the wealth of topological ideas that can be exhibited in an exposition of the sphere eversion. Werner Boy's immersed projective plane, Figure 5:5(11) to 5:5(53), completed the collection of closed surfaces in 3-space illustrating the Kronecker [1869]-Dyck [1888] formula

$$C = 4\chi + 2n\pi ,$$

where C denotes the integral curvature, χ the Euler characteristic and n the number of double curves ending in pinch points. The terms will be defined presently. From Möbius' time it was suspected, though rigorous proofs came later, that every closed surface can be constructed by adding p handles and q crosscaps to a sphere. The alternating sum of the number of faces, edges and vertices of a polyhedral surface is its *Euler characteristic*. For closed surfaces,

$$\chi = F - E + V = 2 - 2p - q .$$

The *genus*, p, uniquely specifies orientable (2-sided) surfaces, such as Riemann surfaces, up to topological equivalence. In this case $q = 0$ because the presence of a Möbius band in the surface would make it 1-sided. The 2-sided surfaces are represented by the familiar sphere S^2, torus T^2, double-torus 7:1(21), 3-holed torus and so forth. All of these can be embedded in 3-space. For non-orientable (closed) surfaces in space you need curves of self intersection. Steiner's Roman surface, 5:3(31), for instance, is the projective plane P^2, for which $p = 0$, $q = 1$ and $n = 3$ by reason of the three double segments. For the usual model of the Klein bottle K^2, we have $p = 0$, $q = 2$; and $n = 0$ because the double curve is closed. Boy's surface was the first example of a P^2 in R^3 without pinch points, $n = 0$.

We have already used the procedure of sewing two bordered surfaces along their boundary curves. The projective plane is the *border sum* of a disc, D^2 and a Möbius band, M^2, as in 5:4(21), and the Klein bottle is the sum of two Möbius bands, 10(23). Two closed surfaces F_1 and F_2, can be joined by removing a disc from each and connecting them along the border circles. The resulting closed surface is topologically independent of the choice of connecting discs in the two surfaces, and it is called their *connected sum* $F_1 \# F_2$. The Euler characteristic is almost additive for this operation, except that 2 must be subtracted for the two discs that are discarded:

$$\chi(F_1 \# F_2) = \chi(F_1) + \chi(F_2) - 2 .$$

The basic combinatorial topology of closed surfaces is summarized by three relations:

$$F = pT^2 \# qP^2$$
$$F \# S^2 = F$$
$$F \# P^2 \# T^2 = F \# P^2 \# K^2 = F \# 3P^2 .$$

I shall convince you of the third relation by showing how $P^2 \# T^2$ is the border sum of a bouquet of three Möbius bands, 2(33), and one disc (not shown). At 2(11) I have drawn the (singular) disc that remains when a Möbius band has been removed from a projective plane. The P^2 is in the form of Steiner's cross cap surface 5:4(21) and a window helps you to see the missing M^2. At 2(21) is a sphere with one handle: the torus. A ribbon neighborhood, H^2, of two circles that cross once has been removed, again leaving a disc. In 2(12) I have broadened the missing M^2 right up to one pinch point. A disc has been removed in preparation for the connected sum. The missing M^2 is at 2(13) and an H^2, missing from a similarly prepared 2(22), is at 2(13). The connected sum is obtained by joining the singular disc, 2(31), to its complement, 2(32). Now shrink 2(32) along the indicated curves (arrows) to produce 2(33). The dotted lines show exactly how this shape is the sum of three Möbius bands. A rewarding, but non-trivial drawing exercise would be for you to modify 2(31) so that the result fits 2(33).

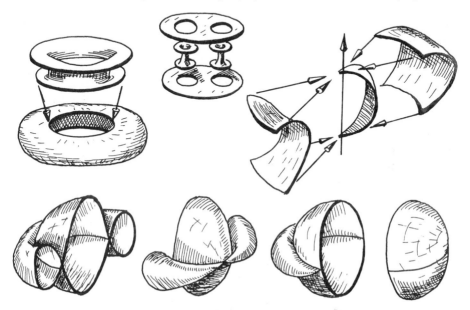

FORBIDDEN EVERSION **Figure 3**

To prove the Kronecker-Dyck formula, Boy had to extend the notion of *integral curvature*, C, to non-orientable surfaces. Gauss' *curvatura integra* is defined only for surfaces that admit a continuous field of normal vectors. For a closed surface, it measures how often the Gauss map, 5:4(11), covers the unit sphere, adding area when the normal is over positively curved portions and subtracting area when it is over negatively curved portions of the surface. The curvature of surfaces generated by (geometric) circles is easy to imagine because you can "transfer" their normals directly to the Gauss sphere, 3(13). Thus, the positively curved outer portion of the standard torus, 3(11), adds one whole spherical area of 4π while the negatively curved *neck* subtracts 4π. Each additional handle, 3(12), subtracts another 4π. The Gauss map is constant on the flat discs with two holes.

In developing his combinatorial version of local curvature density, so that it also makes sense at the singularities (double curves and pinch points), Boy invented *completely continuous deformations* of regular portions of the surface. He considered deformations that might change the self intersection pattern (double locus), as long as the Gauss curvature density remained bounded. At a pinch point the curvature becomes infinite, as you can readily believe when the pinch point is shaped like a cone stuck to a surface, 3(22).

You can turn a sphere inside out easily enough if you do permit pinch points and do not insist on a completely continuous deformation. Imagine a sphere S_1 painted blue on the outside and red on the inside. Now push the North and South Poles towards each other and then right past each other. This forms the immersed sphere S_2, which looks like 3(21) glued to its mirror image. From the outside it looks like a red sphere with a blue tube running around its equator. This form was called the *chapeau* by Morin, who used a hat-shaped version of it, 5(12).

Pinch the tube, as in 3(22), giving birth to a pinch point pair. Now draw the pinch points together on the backside, 3(23), so that they cancel, leaving a sphere with a red outside and blue inside. If you can figure out a way of doing this without the "birth" and "death" of a pinch point pair, you will have, once again, everted the sphere.

That there exist eversions of the sphere is an indirect consequence of an abstract theorem by Stephen Smale [1957]. In principle it is possible to piece together the myriad of minute geometric constructions prescribed by his proof in order to assemble an explicit visualization of an eversion. This strategy is far from practical. In the most recent work on the subject, Homma and Nagase [1984] present a well illustrated piece-wise linear proof of a generalization of Smale's theorem. They meticulously replace Smale's differential topology with a geometric method. But they take intact from the literature the fundamental deformation found by Arnold Shapiro and developed by Bernard Morin, which is pure descriptive topology. The picture story of Shapiro's eversion as remembered by Bernard is told in Francis and Morin [1979].

To understand the wonder Smale's result aroused, and why Raoul Bott insisted on an explicit demonstration of an eversion, let me return, once more, to Boy's dissertation. Applying the notion of completely continuous deformations one dimension lower, to closed curves in the plane, Boy proved that the integral number of times the normal (or directed tangent) of such a curve wound about the circle completely specified the curve up to deformations of the type considered by him. This invariant of closed plane curves has acquired many names since Gauss first dubbed it *amplitudo*. In terms of Whitney's formulation of Graustein's proof [1937] of this theorem, it may be stated as follows. A *regular homotopy* of an immersion is a deformation through immersions for which the matrix of first partials itself varies smoothly. For immersed circles in the plane and spheres in space, Boy's deformations can be parametrized by regular homotopies. The Whitney-Graustein theorem thus says that two immersions of the circle in the plane are regularly homotopic if and only if their *tangent winding numbers* are the same. In contrast, Smale's theorem says that *all* immersions of the sphere in space are regularly homotopic.

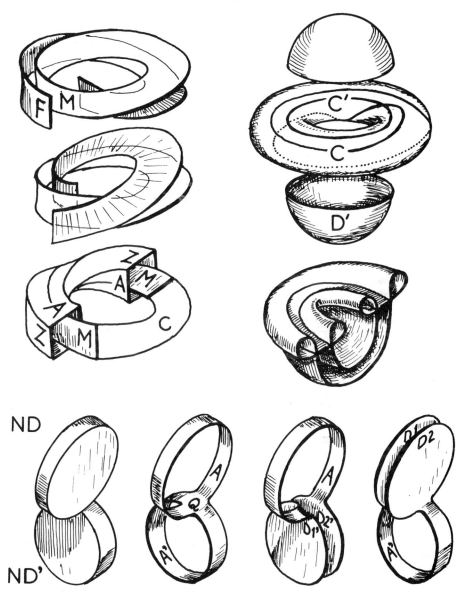

Baseball Move

Figure 4.

The principle of Shapiro's eversion is simple enough, though its implementation is a challenge for descriptive topology. Take Boy's surface B, but split it in two. Think of varnishing its single side and lifting off the coat of varnish in one piece. The resulting surface, call it F, must be an immersed sphere. Here's why. Since B is immersed, it has a continuous field of normal lines, if not of normal unit vectors. If you translate a surface patch uniformly along its normals, it remains uncrumpled, at least until you reach the *focal set* of the patch. This envelope of normal lines is where, in classical terms, "consecutive" normals of a surface meet. For details, consult Chapter 6 of the monograph by Banchoff, Gaffney and McCrory [1982]. Thus, if you push off only a little way from Boy's surface B, then F remains an immersed surface spread out "parallel" to B. Recall that P^2 is doubly covered by antipodal pairs of points on S^2. The normal projection $F \to B$ is thus a physical model of the abstract covering. Let $f:S^2 \to F$ be a parametrization of the surface so that antipodal points on the sphere go to the ends of the segments normal to B.

This splitting is suggested by 4(11), which shows an entire Möbius band flanked by half of its double cover. The other half is shown on 4(21). Now push each point on F uniformly along the normal, first towards B, then right through B as the antipodes pass each other, and then away from B to come to rest again on F. Though the initial and final position of F occupy the same surface in space, the parametrizations differ. The antipodes have exchanged places. This deformation in the target can be parametrized by a regular homotopy from f to the composition of f with the automorphism $S^2 \to S^2$ which exchanges antipodes on the sphere. If you paint the side of F facing B red and the other side blue, then at the end, colors have reversed. Now, all that remains, and that is the hard part to visualize, is to design a deformation of F that removes its self-intersections, yielding an embedded sphere. For then, to assemble the entire eversion, we merely begin with the standard sphere, deform it (by the reverse procedure) to F, evert F through B, and undo F back to S^2.

Shapiro never drew a picture of F. He argued topologically as follows. Isolate a "standard" Möbius band M on B, which is embedded with one half-twist in space. Can you find one of these on 5:5(13)? Thicken B to a thin solid NB and carve out the corresponding neighborhood NM of M, see 4(31). The surface $T = \partial NM$ of this solid is a topological torus with corners. It is composed of two ruled annuli joined at right angles along their borders. One ribbon A crosses the border curve $C = \partial M$ of the Möbius band at right angles. The other ribbon $Z = T - A$ is parallel to M. The complement of NM in NB is an immersed solid disc ND that intersects itself and NB in a most horrible way. No matter! ND is the immersed image of a thin cylinder, call it a *coin*, with two face discs, $D1$, $D2$, parallel to D, and an edge ribbon, which is A, of course.

$$\begin{aligned}\partial ND &= D1 + A + D2 \\ \partial NM &= \phantom{D1 +{}} A \phantom{{}+ D2} + Z = T \\ \partial NB &= D1 \phantom{{}+ A} + D2 + Z = F.\end{aligned}$$

Now girdle T by any simple closed curve C' that crosses C exactly once, for example a longitude on the torus, 4(12). The distinction between *meridian* and *longitude* on a toroidal tube (= thick knot) embedded in space is useful in knot theory, as we shall see in the last chapter. A meridian bounds a disc on the inside of the tube, a longitude bounds discs on the outside. Consider the longitude C' as drawn on top of the torus in 4(22) and span it by the upper hemispherical cap. This topological disc spans the longitude on the outside. Complete the hemisphere to a sphere. The lower hemisphere D' also spans C', but not on the outside, because is passes through the torus. It is this disc which is to be attached to the torus along C' and thickened to produce another coin with boundary $\partial ND' = D1' + A' + D2'$. The edge ribbon A' of this second coin crosses A at a little square Q. Glue the two annuli together (the operation is called *plumbing*) to form an immersed Seifert surface $H = (A - Q) + Q + (A' - Q)$. A model of the two plumbed ribbons is shown in 4(42). The complement of A' on the torus is another annulus, call it Z'. Note that the surface $F' = D1' + D2' + Z'$, is also an immersed sphere, but this one is easier to visualize; it has no triple points. Half of it is shown in 4(32).

Shapiro invented a certain regular homotopy, known as his *baseball move*, which takes F to F'. To understand this regular homotopy we look at the source of the mappings. The union of ND and ND' is an immersion in R^3 of a solid that looks like two coins welded along a flattened square Q on their edges, 4(41). (I shall use the same letters to denote surface patches on the model and in space.) A wire *hoop* along the "corners" of the welded coins takes on the shape of a baseball seam if you inflate the skin of the welded coins into a sphere. The hoop spans the two discs $D1'$ and $D2'$, connected by a band $A - Q$, to form one topological disc $E' = D1' + D2' + (A - Q)$, in the shape of Morin's *earphones* 4(43). The same hoop spans a second set of earphones $E = D1 + D2 + (A' - Q)$ and there is an deformation of 4(43) to 4(44) which occurs entirely within the space of the two welded coins. Chop down on A with the edge of your flat hand, pushing the surface against the faces of the upper coin. Midway is the saddle $D1 + D2 + Q + D1' + D2'$. Now push Q down, peeling the surface off the lower discs into the band A' of E. On Shapiro's baseball, push one flap through the ball to the other. In the model, the deformation of E to E' is an isotopy, its image in space it is a regular homotopy. Note that most of $Z = T - A = (A' - Q) + (T - H)$ remains stationary as F moves to F' as summarized by the following bit of algebraic topology:

$$\begin{aligned} F &= D1 + D2 + Z \\ &= D1 + D2 + (A' - Q) + (T - H) \\ &= E + (T - H) \cong E' + (T - H) \\ &= D1' + D2' + (A - Q) + (T - H) \\ &= D1' + D2' + Z' = F' . \end{aligned}$$

Shapiro needs another baseball move that takes F' to the surface F'' obtained by pairing the longitude to a meridian. I leave to you the task of illustrating this baseball move and the final uncrossing of F'' into an embedded sphere. You can find one solution to this exercise of descriptive topology in Francis and Morin [1979].

This foregoing synopsis of Shapiro's idea was written in the descriptive style favored by topologists, though customarily it comes with fewer pictures and more precise coding of the details. Illustrations suitable for an article in *Scientific American* have to be simple and leave less to the untrained imagination of the viewer. Tony Phillips [1966] designed ingenious pictures of several stages in a regular homotopy between the standard sphere and the double cover of Boy's surface. Unfortunately, the magazine's artist introduced a number of graphical errors in his rendering of Phillips' plan. These distract the eye as you try to comprehend each depicted surface and imagine the transitions between them. The same misfortune befell Jean-Pierre Petit's superb designs for the definitive publication of Morin's eversion [1979]. Though better geometrical training for graphical artists would have led to happier collaboration, mathematical pictures are notoriously difficult to proof-read. I redrew the final version of the pictures for this book myself, and yet I am sure that you will find errors that inserted themselves in the course of copying the drafts or even earlier.

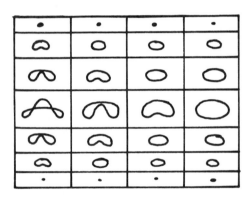
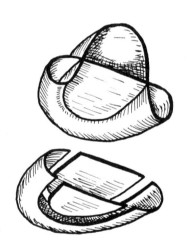

Chapeau Diagram

Figure 5

It is more interesting and instructive to compare the topological design of the eversions drawn by Phillips and Petit. The former is a mathematician, the latter a physicist, and both are good artists. Phillips used the sections of Boy's surface by parallel planes and doubled them to obtain sections for F. Special care is taken at the saddle of B. To suggest a surface corresponding to a sequence of plane diagrams, he connected spatially consecutive curves by surface *zones*, similar to those of 5:7(13) to (53) but narrower. The zones are stacked vertically, with space between them for visibility. Corresponding temporally consecutive zones are drawn on the same horizontal level. This technique enables you to work with a 2-dimensional array of plane diagrams, which makes it easy to work out the topology of the regular homotopy. The columns are the floor plans for the storeys, the rows show their deformations in the earthquake rocking the tower. For example, 5(11) is such a diagram for the regular homotopy that uncrosses the chapeau, 5(12), which was used in the singular eversion earlier. The chief problem with Phillips' method is that the slicing direction, once chosen for B, remains fixed. The 2-way diagrams are not isotropic; rotation of the slicing direction produces unrecognizably different diagrams. A secondary problem is the complexity of the surfaces and their motions, once the diagram unfolds into a picture. The self-intersections, mere points in the diagram, become strangely twisted curves in space. Petit's drawings, on the other hand, are really pictures of Morin's architectural plans for 3-dimensional constructions. Once you comprehend the object depicted in each detail, you would recognize it from any angle, even if you stepped inside one of its chambers.

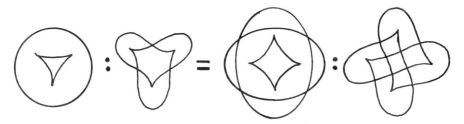

Astroid/Deltoid

Figure 6

Bernard Morin is not distracted, like the rest of us, by pencil and paper and the business of drawing and looking at pictures. He is blind. With superb spatial imagination, he assembles complicated homotopies of surfaces directly in space. He keeps track of the temporal changes in the double curves and the surface patches spanning them. His instructions to the artist consist of a vivid description of the model in his mind. Petit, in effect, draws the storyboard for an animated film. Nelson Max [1977] realized just such an animation with the aid of spectacular computer graphics. He used the coordinates of Charles Pugh's meticulously soldered wire models as a database. It was only when I guided my 1977 freshman topology seminar through Max's film, sketching on the blackboard to analyze the motion, that a disadvantage of this 3-dimensional approach became plain. It is beyond the ordinary artistic means of the descriptive topologist to reproduce, modify, dissect and otherwise manipulate the details of such an eversion. In the course of the seminar, we noticed that the top-view contour of Morin's surface, 6(11), bears the same relation to the astroid, 6(12), as that of Boy's surface, 6(13), bears to the deltoid, 6(14). With this discovery the designs for the tobacco pouch eversions were on the way and the idea of this picturebook was conceived, see Francis [1980].

Assembling Immersions. *Figure 7.*

With graphical techniques that I had developed over the years to help me prove theorems about mapping surfaces to the sphere, I could construct a picture of Morin's surface, 8(33), beginning from its astroid shaped contour, 6(12). You can ignore the two crossed ovals, which correspond to the horizons of two ellipsoidal shells observed when looking at the surface from the outside. If you step inside and project the surface radially to a picture sphere, then only the astroid contour remains. Since it is difficult to think of drawing pictures on a sphere, I shall continue to refer to a picture plane, with the understanding that any surface drawn with a given contour has sheets that continue indefinitely and eventually parallel to the picture plane, insofar as it does not fit completely inside the frame of the picture.

Let me give you an intuitive idea for this technique, without bothering with the mathematical details. These, together with their history, are explained in Francis-Troyer [1977]. If, like a flower, you press such a surface flat against the picture plane, the contour lines become fold lines. Unfold the surface and note how the creases and their traces on other parts of the surface divide it up into regions. Each region, without its boundary, is embedded in the picture plane. Such a region need not be a topological disc. Flattening the top view of a torus produces two annuli. If, however, you had laid out enough arcs across the picture, and copied their traces on the flattened surface, all the regions could be discs. One arc crossing both contours of the torus will do here. These discs can serve as (provisional) windows in the picture of the surface. In 7(11) I have widened and separated two such arcs which cross the annuli in an unexpected way. Shrink the windows, 7(12), and observe that I have captured an immersed torus instead of the embedded one, which is generated by a figure-8 spun about the vertical axis.

There is a combinatorial machine which tells you how to lay out the arcs, identify the windows, and assemble pieces into an abstract surface and a concrete mapping of it into the picture plane, which folds along the given curves. It also classifies all possible ways of doing this, up to a topological equivalence on the source surface. Because of its great generality, and logical rigor, the method is rather impractical, though available in a pinch. In practice, draw a right- or left-handed cusp neighborhood at each cusp of the given contour, and connect them up, with or without a pinch point on each contour. With a little practice, you can spot embedded surface pieces spanning the borders of the bent and cusped ribbons so produced, and divide them up into windows. The seven steps from 7(21) to 7(23) show how 5:8(53) came from 5:8(54).

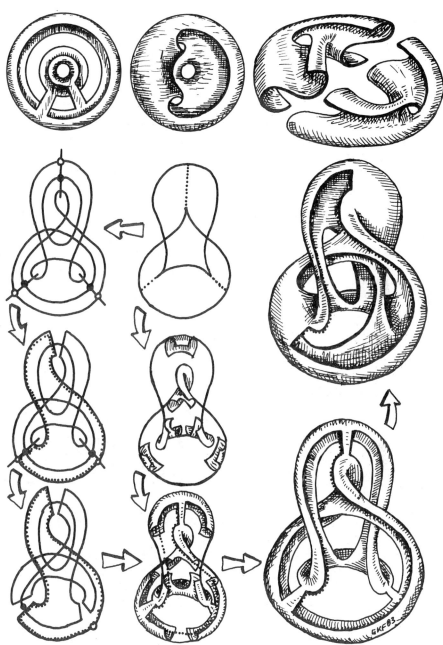

Figure 7 ASSEMBLING IMMERSIONS

MORIN EVERSION. *Figure 8.*

The bent ribbon in 8(22), with four left-handed cusps, is an annulus, immersed in R^3 with four half-twists. So is 8(21), but its contour is not an astroid. The motion from 8(21) to 8(22) is clear: just push in the direction of the arrows, keeping the cusps fixed. Since no pinch points are produced, this is a regular homotopy of the ribbon. Now draw two narrow oval annuli around the two figures. (You will need tracing paper and pencil to follow me.) There are many ways of connecting the cusps by four strips to the oval rims. Some ways do not produce windows. In 8(11) and 8(12) I have chosen some that do. Moving the contours merely stretches or shrinks the windows, but does not, for example, bend a window over far enough to produce new contours, or worse. Whichever way you close the windows in 8(11), the motion of their frame pulls them right along to some closure of the windows of 8(12). There is no unique position of the windows normal to the picture plane. But you can close the windows so that the resulting surface is an embedded annulus in the shape of a mildly twisted torus neck. Since the borders of the windows in 8(12) link, whatever position the windows are in, there will be double curves. It is not easy to draw even the simplest version. That is another good reason to use windows. In principle you can avoid drawing the double locus entirely with the help of windows. But sometimes the window rim crosses over the contour, as at pinch points. See 5:4(21) for example. Such windows are not as mobile.

Now push the nearer oval rim of 8(12) down into the picture plane, right through the further rim. Call it picture 8(12') because you could xerox 8(12) and turn it 90°. Of course, your "holey" torus, 8(12'), will have the colors blue and red reversed. On the toroidal neck, with oval borders made circular, you will have created a circle of self intersection C, concentric to the borders. Next pull the vertical arms of 8(22) horizontally across each other. The motion is the same as the one from 8(21) to 8(22), but turned 90°. This takes 8(12') to 8(11'), which is 8(11) turned through 90°, but the colors have reversed. If you imagine the toroidal neck firmly attached to two concentric spherical shells, 8(31), then the surface so obtained is an upside down chapeau, 7(12). Uncrossing the chapeau completes an eversion of the sphere in the style of Morin. Any two ways of closing a window by specifying the third coordinate are regularly homotopic by a *back and forth* motion that leaves the projection to the picture plane invariant. I shall only be interested in the circle C of the chapeau stage.

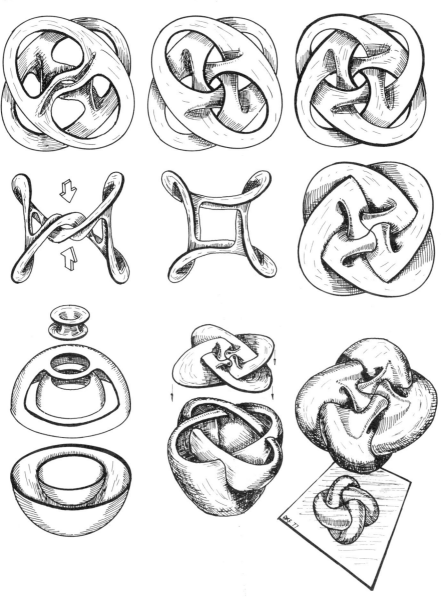

Figure 8 Morin Eversion

Recall that I asked you to create C all at once between 8(12) and 8(12'). If, instead of pushing down on all four places on the nearer rim, you do so only at two, as in 8(13), the result is a shape with 4-fold rotational symmetry. Closing these windows, 8(23), produces an immersed annulus, which fits into the two crossed, ellipsoidal shells, 8(32). When their South Poles coincide, the crossed shells and crossed neck form Morin's surface, 8(33). Morin's surface everts by rotating it 90° about its symmetry axis. Were you to produce an animated film of the eversion, like Max's, you could use the trick of photographing the frames in reverse order and reverse color.

TOBACCO POUCHES. *Figure 9.*

The aesthetically pleasing symmetry of the middle stage of Morin's eversion generalizes to a sequence of them, named after the common French *tobacco pouch*, 9(11) and 9(12). This homely rubber object has a number (usually five or six) of creases which allows the pouch to close automatically, as the rim turns against the bulb. If the material could penetrate itself, the motion would continue as a deformation that turns the pouch inside out. The principle is always the same. First twist $n = 2,3,4,...$ *swallowtails* into the circular contour of the neck, like the three in 9(31). (The name swallowtail is borrowed from Morin's designation for the characteristic shape associated with one of the elementary catastrophes.) I have drawn cases $n = 3$ and 5 as polygonal knot diagrams, 9(21) and 9(23). Can you figure out the overpasses in 9(22)? The small, unkotted curves deform into the larger knots around them by a regular homotopy in space whose shadow shows each edge of the unknot stretching as it moves parallel to itself. The longer edges move towards the center, and come to rest on the far side after crossing each other all at once in the middle. The shorter edges move away from the center a little, but stretch a lot. Note that in the odd cases, the polygonal knot can continue its motion until it runs around a cable knot twice, as I have drawn in 9(23). Now treat the curves as contours of a bent ribbon with all cusps of the same kind. For example, the ribbon 9(31) corresponds to the unknot in 9(21), and 9(32) to the trefoil knot in 9(21). The homotopy of the contour extends to the ribbon. In the even case, the ribbon is 2-sided, and changes color under a $1/2n$ turn. In the odd case, the ribbon doubly covers a Möbius band with n half-twists. The bottom row shows how the ribbon for the case $n = 3$ eventually folds up to a Möbius band with three cusps. You can untwist the cusps to check this, 9(34). Thus, once the regular homotopy so coded is extended to a sphere, initially shaped like a *gastrula*, 8(31), the eversion passes through a general Boy's surface, like 9(13) for $n = 5$.

The design of these regular homotopies proceeds in three steps. First, determine the motion of the contour in the picture plane. Next, *elevate* this motion into the third dimension as follows. Grow the contour into a ribbon

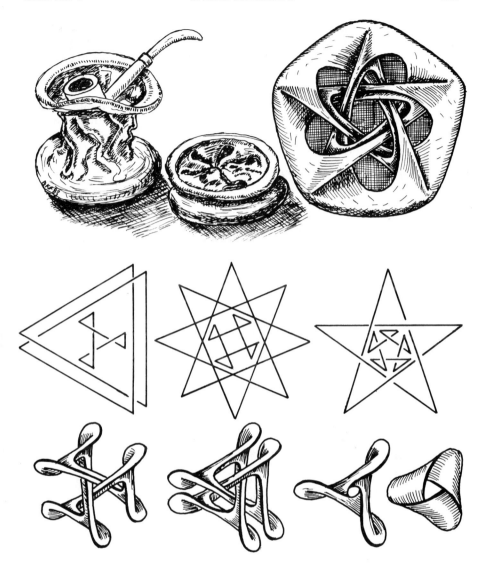

Figure 9 TOBACCO POUCHES

in a manner independent of the deformation. That way, the moving contour in the plane is elevated into a moving ribbon in space. Finally, show how the motion of the ribbon extends to a regular homotopy of the entire surface. In an elegant piece of analytic geometry, Morin parametrized just such a procedure with a set of trigonometric expressions [1978]. To illustrate how this works, I shall sketch the first two stages in a somewhat simplified form. The more technical globalization is beyond this story. Moreover, for the third stage the graphical and analytical approaches do not correspond very well.

For each $n = 2,3,...$ and direction $0 \leq \theta \leq \pi$ on the circle S^1 define a hypocycloid Γ in the (complex) plane. Together with first and second derivatives, its definition is:

$$\Gamma = e^{-(n-1)i\theta} + \frac{n-1}{n+1} e^{(n+1)i\theta}$$
$$\Gamma_\theta = -2(n-1) \sin(n\theta) \, e^{i\theta}$$
$$\Gamma_{\theta\theta} = -2(n-1) \{n \cos(n\theta) - i \sin(n\theta)\} \, e^{i\theta}.$$

Note that its tangent line at θ has direction $e^{i\theta}$. This line turns continuously through $2n$ semi-cubical cusps, because the acceleration is not zero when the velocity is. For odd $n = 2k + 1$, the parameter covers its track twice. The track is an astroid for $n = 2$, a deltoid for $n = 3$ and a cyclic star thereafter. Expressing the curve as the envelope of its tangents yields a map from the cylinder to the plane which folds along the curve. Here are the algebraic details:

$$R^1 \times S^1 \to R^2 : (r,\theta) \to L = \Gamma + r \, e^{i\theta}$$
$$L_r = e^{i\theta}$$
$$L_\theta = -2(n-1) \sin(n\theta) \, e^{i\theta} + r \, i e^{i\theta}.$$

The determinant of these last two vectors is the Jacobian of the map. Since it equals r we have shown that the map is excellent in the sense of Whitney. To the third coordinate, perpendicular to the picture plane, assign the height function

$$H = \frac{n-1}{2n} \sin(2n\theta) + r \sin(n\theta)$$
$$= \left\{ \frac{n-1}{n} \cos(n\theta) + r \right\} \sin(n\theta).$$

This makes the space curve (Γ,H), whose shadow is the star, undulate across the picture plane. It also rotates the lines L in planes normal to the picture plane, except at the $2n$ cusps where $\sin(n\theta) = 0$ and L remains horizontal. The elevated function $F = (L,H)$ maps the cylinder to a ruled surface in space. It is regular (an immersion) because r is the third component of its normal vector $N = F_r \times F_\theta$, and when $r = 0$, $N = (n-1)ie^{i\theta}$. Its contour is the cyclic star.

It is remarkable that the star is the midstage, $t = \frac{1}{2}$, of the deformation which slides each line in the tangent envelope of a circle of radius $(n - 1)$ to the opposite point on the circle, all the while remaining parallel to itself. Use trigonometric sliding coefficients $A = \cos(\pi t)$, $B = \sin(\pi t)$ and compute:

$$C = (n - 1)ie^{i\theta} A + B \, \Gamma$$
$$C_\theta = -(n - 1)\{A + 2B \sin(n\theta)\} \, e^{i\theta}$$

To elevate this motion to a regular homotopy of the space ribbon, proceed as before. It is, however, necessary to adjust the r-parameter by a positive factor $M = \sin(\pi t) + 0.2$, for example, to insure that for all $0 \le t \le 1$ the 2 by 3 Jacobian matrix of the mapping

$$W = C + Mr \, e^{i\theta}$$
$$Z = \{r + 3 \, \frac{n-1}{n} \cos(n\theta)\}A + BH$$

has maximal rank at (hence near) $r = 0$. You should practice your calculus by checking this out and pursuing Morin's analysis further. What has been accomplished so far is the eversion (up to a rotation by π/n) of a wobbly tobacco pouch neck, whose contour is C.

ASTROID FAMILY. *Figure 10.*

If you are used to mathematical problems that have only one tidy solution, you may wonder about the variety of ways Bott's challenge has been met by descriptive topologists, and ask how many different ways one can design an eversion of the sphere. To answer your question would depend, first of all, under what circumstances you would agree that two eversions are "the same". Just as taxonomy is the wellspring of biology, the classification of topological species, up to an equivalence relation, has always played a central role in topology. I cannot tell you much about classifying eversions. That is too difficult. Instead, let me pursue the more practical matter of growing surfaces from contours a little further, and introduce you to the other members of the astroid family. The rules of the game were laid down earlier. Find mappings $W = \Phi(w)$ from borderless surfaces to the picture plane which fold only along the astroid, and construct elevations $Z = H(w)$ so that $F(w) = (\Phi(w), H(w))$ parametrizes normal surfaces with astroid contours. Φ may branch only "at infinity," so that any pinch points in F are on the contour. Two "solutions" are deemed equivalent if their projections differ by an automorphism of the source. In Francis and Troyer [1977] we found six such Φ. The seventh, an excellent mapping of the Klein bottle, eluded us for quite a while, see Francis and Troyer [1982].

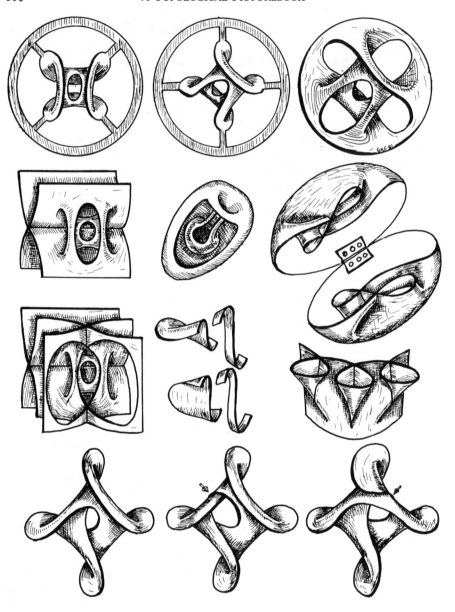

Figure 10 ASTROID FAMILY

Suppose that, instead of drawing all cusps the same way, you alternate their parity, *LRLR*, as in 10(11), or change once, *LLRR*, as in 10(12). Let us leave pinch points till later. Both ribbons are embedded annuli with no twists. You can unbend them. Suppose, further, you want the surface to continue as one sheet beyond the edge of the picture. The four little bridges establish three windows. Shrink the windows of 10(11) to reveal the handle for a torus. Shrink those of 10(12) for the handle of a Klein bottle, 10(13). If you complete the latter to a closed suface by gluing it to a bowl, you can split the Klein bottle into two (immersed) Möbius bands, 10(23).

There is only one way to build bridges so as to continue on two sheets. For *LRLR*, this yields 10(21). The corresponding closed surface is an immersed sphere. Can you draw it? For *LLRR*, you obtain the embedded sphere, the gastrula, 10(22), also with two windows. I have reproduced all three stages of construction only for the Klein bottle, 10(12), 10(13) and 10(23). For practice, you should try to draw the missing stages for the other six cases.

All ways of attaching three or more rims to these ribbons fail to yield windows, except one, 10(31). This immersion of the plane in Euclidean 3-space is isotopic to *Enneper's surface*, which minimizes area, and has an astroid contour when seen from above. Its Weierstrass-Henneberg normal form, as discussed in Nitsche [1975], may be expressed in mixed coordinates, $W = X + iY$, $w = x + iy$, thus

$$W = \overline{w} - \tfrac{1}{3}w^3$$
$$Z = Re(w^2) \ .$$

The projection to the *W*-plane has a triple pole, as illustrated in 10(31) by the single, outside border, which winds three times about the picture. At the origin, the surface map stabilizes the singularity of the complex cubing function, by folding along the astroid. This contour is the image of the unit circle, $w = e^{i\theta}$, which maps to the hypocycloid,

$$W = e^{-i\theta} - \tfrac{1}{3}e^{3i\theta}$$

The ribbon 10(41), with cusps reading *RLRR* clockwise from noon, is an annulus embedded in space with one twist. You can check this by unbending the contour between 3 and 6 o'clock, and untwisting the two remaining ears with the graphical recipe shown in 10(32). It also follows from the fact that the two borders of 10(41) link once. Such a ribbon is ineligible because you cannot continue the borders to infinity without introducing new folds. This is also a consequence of a theorem about closed, 2-sided ribbons embedded in space, knotted or not. Such a ribbon, let me call it a *Morin band* in analogy to the Möbius band, can be sewn to an immersed disc (on either edge) if and only if the two edges link an even number of times. Even and odd Morin bands are very useful in descriptive topology and this folk theorem has a simple picture proof, but that is another story.

That brings us to the case of pinch points. You can slide a pinch point past a cusp on its contour, 10(33), switching the parity of the cusp on the ribbon. Pinch points adjacent on a contour segment cancel pairwise, as in 2:9-11, till at most one is left. This leads to only two cases. I have drawn representatives for them at 10(42) and (43), reading *RLRRP* and *RPLRR* respectively. The borders of these Möbius bands have been shrunk to cross the contour at the pinch point, arrow. 10(42) has three half-twists and the ears untwist to the mirror image of 10(34). 10(41) is a standard Möbius band with half a twist. Both extend to infinity, but only 10(42) does so without additional folds. Can you complete this to a drawing of a cross cap with an astroid contour?

CATASTROPHE MACHINE. *Figure 11.*

As mentioned before, I first met Morin's surface disguised as a member of the astroid family in connection with singularity theory. Drawn as in Figure 11, the projection of Morin's surface gives a simple, topological model for Zeeman's celebrated catastrophe machine [1972]. Two rubber bands connect a pin on the freely rotating flywheel to a fixed pin on one side and a movable point, say a pencil, on the other. As you move the pencil about the *control plane*, the wheel jumps in mysterious ways. You can predict the jumps by lifting the path of the pencil to the surface. Minima of the machine's potential energy are represented by the points on the sheet drawn, including the two windows, up to the contour. Color this sheet blue. The potential maxima, which the machine avoids, are located on the other, red, half of the surface. I drew only a narrow portion of the red sheet near the contour. The wheel jumps at discontinuities of the lifted path. For example, a tour of the pencil clear around the astroid is jumpfree because its lift stays continuously on the blue sheet. During a horizontal pass across the astroid, the wheel jumps as the pencil leaves the region. This is so because the lifted path emerges from one window and falls off the contour to the other window. Vertical passes through the region inside the astroid, which don't cross the center line, don't cause jumps because the lift merely crosses on a window. It was this and similar catastrophes, regarded from the viewpoint of descriptive topology, that originally motivated my interest in classifying excellent mappings of surfaces in terms of their contours.

MORIN TWIST. *Figure 12.*

Let us have another look at Morin's tobacco pouch eversion for the case $n = 2$. Its anatomy consists of the annular rim, the creased neck and the bulb, which contains the tobacco. You twist the rim against the bulb to get at the tobacco. The pouch closes automatically, twisting the other way.

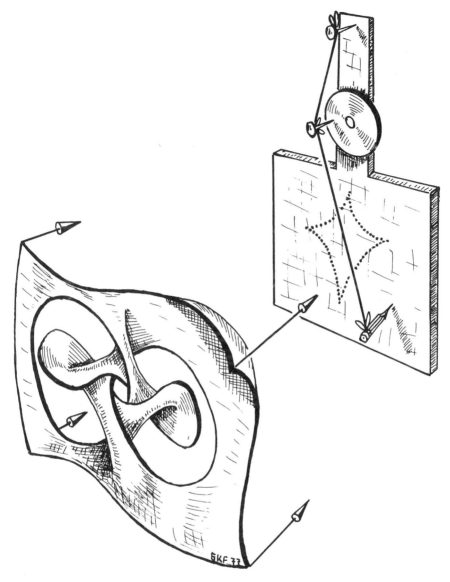

Figure 11 CATASTROPHE MACHINE

Continue this motion to move the neck through itself. The rim and the base of the bulb never leave their respective horizontal planes, and the neck never moves outside a cylindrical space between them, call it a *pill box*. By the end of this deformation, the neck crosses itself along a circle of double points concentric to its borders. Since the pouch is made of rubber, you can hold the rim and bulb fast, so that no point on them moves at all. In fact, the base disc of the bulb is irrelevant; you have deformed the neck and kept its two rims pointwise fixed.

Since the rims remain fixed in space, you cannot uncross the neck as in the deformation of the chapeau. Instead, repeat the motion, but draw the cusps with opposite parity. In the middle of the figure, 12(22), I have marked the neck with the initial and final cross cut. At 9 o'clock is a stylized drawing of the starting position, with very large windows and the final position of the cross cut. The smaller rim is nearer than the larger one. This *abus de dessin*, sometimes called *Chinese perspective*, in which an object shrinks as it comes nearer, makes it easier to design and read complicated diagrams. Follow the fate of the cross cut on the figures read clockwise. The neck is closing at 11 o'clock, by means of left handed cusps. It reaches Morin's midstage by noon, and re-opens at one o'clock. Note that the windows at 3 o'clock link, so the neck is immersed, not embedded. Now twist the neck again, but with right handed cusps this time. It definitely helps if you color the sides of the details from 3 to 9 o'clock. Note that 6 o'clock is a mirror image of 8(12) not of 8(13). By 9 o'clock the neck is in its original position but, as the cross cut shows, it has experienced a *Dehn twist*.

This is an automorphism of an annulus that maps concentric circles into themselves, keeping the points near the border fixed. Each radial segment, called a *cross cut*, however, is twisted once about the hole in the annulus. It is not possible to twist an annulus by a deformation in its plane, which fixes the borders. It is usually described in terms of a cut, twist and paste operation. Here, however, you have seen that a deformation inside the pill box produces a twist without cutting and pasting. Self maps of surfaces can often be visualized by deforming a particular model of it in space, which returns to its original shape. Since you cannot really follow each point to its new position, certain auxilliary curves are marked on the surface and kept track of during the motion. That trick is used in the next two picture stories.

Let me end by mentioning two stories told by other people, where Morin's tobacco pouch move plays a central role. Like Shapiro's baseball move, it is a model that can be grafted into a given surface, F. Modify F by an immersion of the pill box, that takes its cylindrical face to an even Morin ribbon M on F, and its top and bottom parallel to an immersed disc spanning M. One Morin move produces a regular homotopy from F to a modification F' in which M is replaced by a toroidal tube that crosses itself along the centerline of M. A second Morin move, of opposite parity, removes this double curve, replacing M by one Dehn twist of M. The insertion and removal of double curves that fit Morin's move is a basic device in Nagase's construction [1984]. He shows how any immersion of the disc in space, with its border anchored to a plane annulus, is regularly homotopic to the disc in the plane of its border. This was the main lemma of Smale's invisible theorem as well.

Figure 12   Morin Twist

Note that a Dehn twist put into a spherical zone by Morin's regular homotopy can be untwisted by an isotopy in one of its hemispheres. Thus we have a closed loop in the space of smooth immersions of S^2 in R^3. Morin's conjecture that this loop represents the generator of the infinite cyclic fundamental group of this space was proved by Max and Banchoff [1981]. This paper contains a good example of a "picture proof." This proof relies on hand drawn sketches of the computer designed views of the moving surface in Max's film [1977].

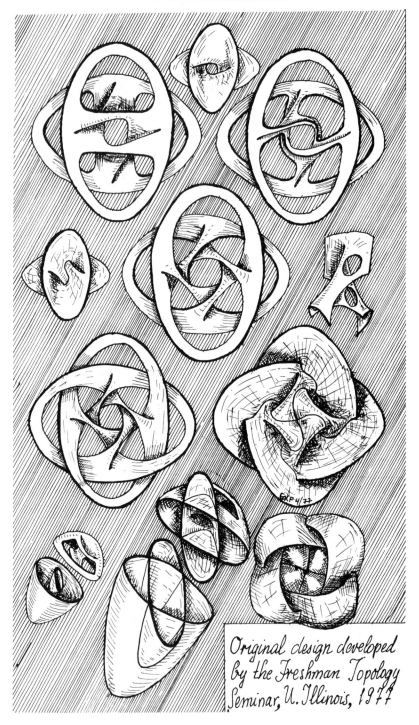

Figure 13

7
GROUP PICTURES

There are two important groups associated with a closed surface: the fundamental group and the mapping class group. The former, also known as the first homotopy group of the surface (considered as a topological space), is probably the more familiar of the two. It was defined and explored by Henri Poincaré [1895]. In this chapter I would like to revive a few of the traditional images that go with the second, more picturesque group. Customary definitions of these two groups go something like this.

For the *fundamental group*, choose a *base point* on the surface and consider the set of all continuous paths that begin and end at this point. The possibility of deforming one such loop into another establishes an equivalence relation which respects the binary operation on loops defined as follows. The *product* of two loops is obtained by joining the end of the first to the beginning of the second. For a tour of the surface on the product loop you must leave for the second trip immediately on returning from the first. The *identity element* of the resulting group is represented by loops that can be pulled into a point like a lasso. For example, if you turn around at the end of a tour and retrace your path backwards, this product of a loop and its *inverse* can be pulled into a point without ever leaving the path. More generally, a *null homotopic* or *trivial* loop may sweep across a patch of the surface as it is pulled into the base point.

On a sphere, for example, there are no handles to tie a lasso around, and the fundamental group is itself trivial. Every other surface has at least one handle or cross cap which catches a nontrivial member of the fundamental group. Poincaré conjectured that the fundamental group, together with its higher dimensional analogs, always has this ability to distinguish spheres from more complicated manifolds. Testing this question has been a mainspring for topology for nearly a century. The conjecture was confirmed by Steve Smale [1960], but only for dimensions greater than four, and by M. Freedman [1980] for dimension four. For our world of three dimensions its truth remains unproven.

For the *mapping class group*, consider the set of homeomorphisms of the surface to itself (the self-maps). In this context, a *self-map* is always a one-to-one transformation of a closed, orientable surface onto itself which, together with its inverse, is continuous. The possibility of deforming one self-map into another through a continuous family of self-maps, called an *isotopy*, establishes an equivalence relation which respects composition of mappings. The *product* of two self-maps is just the self-map obtained by performing the second immediately after the first.

You may think of the fundamental group of a surface as an algebraic record of the difficulties you encounter in trying to pull loops to the base point. Similarly, its mapping class group is an algebraic record of the obstructions to deforming self-maps to the identity map. The *identity map* takes each point of the surface to itself. The fundamental relation between these two groups was first recognized by Jakob Nielsen [1927]. In particular, the mapping class group is isomorphic to the group of outer automorphisms of the fundamental group of the surface. Since I won't pursue this branch of the theory I must refer you to the literature for further detail.

ZIPPING UP A DOUBLE TORUS. *Figure 1.*

To contemplate the self-maps of a surface requires first of all a good way of imagining the surface placed in space. For the *double torus*, which is the closed orientable surface of genus two (Doppelring), take an octagon, 1(11), and glue the edges pairwise as shown in 1(12). Note that edge 1-2, from vertex 1 to vertex 2, is glued to edge *43*, which is *34* in the reverse direction. If you use alphabetic labels for the edges, so that A = 1-2 and A^{-1} = 3-4 etc., then the octagon reads

$$A \ B \ A^{-1} B^{-1} C \ D \ C^{-1} D^{-1} .$$

Note that the resulting cell complex has only one vertex. Hence the four edges form a bouquet of loops, 1(13). Traversing these loops as directed by the above word takes you around the original octagonal disc exactly once. Hence this concatenated loop is null-homotopic and represents the identity. Indeed, this is the only relation needed to give the canonical presentation of the fundamental group of the double torus.

Let us look at this procedure as it is applied to the simple torus. Identifying horizontal and vertical edges of a rectangle produces a surface you can experience on a wrap-around computer screen. Figure 1(31) shows two cylinders produced by zipping up the vertical edges first. The tall tube deforms smoothly into the narrow band. Now sew up the two border circles of each surface to obtain two linked tori, 1(32). These two closed surfaces are subtly different from each other. If you turn the horizontal torus 90° and move it back a little, it will occupy the same place as the vertical torus. Such a *superposition* of two copies of the same surface defines a self-map.

CHAPTER 7 GROUP PICTURES 127

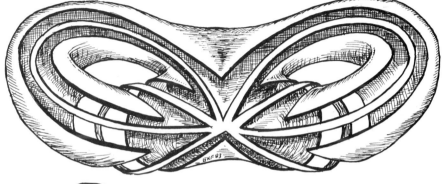

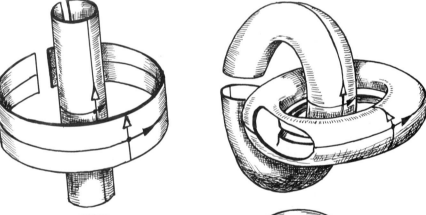

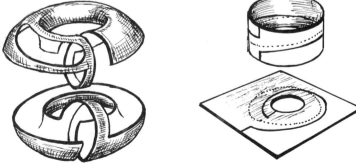

Figure 1 ZIPPING UP A DOUBLE TORUS

Actually, it defines the self-map only up to an isotopy on the surface and hence it corresponds to an element of the mapping class group.

This particular element of the mapping class group of the torus is not the identity, it has order four. To see why this should be so you can follow the effect of the self-map on certain circles on the torus. (This being topology, call any simple closed curve on a surface a *circle*.) A closed surface embedded in 3-space separates space into the inside and the outside of the surface. A circle on the surface which is the border of some disc embedded in space is said to be *spanned* by the disc. (Such a circle is *a fortiori* unknotted.) A circle which is spanned by a disc lying in the outside of the surface, as well as by a disc on the inside, is spanned by a disc on the surface and is therefore homotopically trivial. A nontrivial circle is called a *meridian* (resp. *longitude*) if it is spanned by a disc on the inside (resp. outside) only. Up to an isotopy, a torus has one meridian and one longitude. It also has unknotted, nontrivial circles which are neither. Can you find them? A self-map represented by a superposition of the torus must, of course, send meridian to meridian and longitude to longitude.

Now imagine the self-map of the vertical torus in 1(32) thus. Cut it open along a meridian, straighten it out to the tall cylindrical tube in 1(31). Shrink the height while you expand the diameter of the tube and curl the band of 1(31) inward to produce the horizontal torus of 1(32). Turn this 90° counter clockwise and place it on top of the original surface. This self-map, call it the *swap*, takes the meridian (black arrow head) to the longitude (white arrow head) and the longitude to the inverse of the meridian. The swap has order four. I have described the swap in a way that generalizes neatly, as you will see below.

There is another way of modifying a torus which has been cut open along a meridian. Twist the tube a full turn and then sew it up again. To see what happens under such a *twist*, imagine the horizontal torus composed of two annuli sewn together along the two equatorial longitudes. Now rearrange the annuli in space as shown in 1(41), and move them together to form a horizontal torus. This self-map takes a longitude into a curve composed of one longitude plus one meridian. (This curve solves the puzzle posed above.) Twist again and the curve picks up another meridian. Iterating this self-map never returns to the identity.

The square of the swap takes meridian and longitude into their inverses. The superposition for this self-map requires no cutting and pasting, just turn the vertical torus 180°. Thus it commutes with the swap. That the product of swap and twist has order 3 is much harder to see and I shall return to this at the end of the chapter. Thus we have one presentation for the mapping class group of the torus:

$$\langle s,t \mid s^4 = s^2ts^{-2}t^{-1} = (st)^3 = 1 \rangle.$$

Of course, the true identity of this group is unmasked algebraically in terms of the abstract definition. Nielsen thought of the mapping class group as the automorphism group of the fundamental group of the surface. Since the rectangle gluing recipe is $ABA^{-1}B^{-1}$, the fundamental group of the torus must be the free Abelian group Z^2 of rank two. The automorphisms of Z^2 comprise $GL(2,Z)$, the general linear group of two by two matrices over the integers. As you can readily check, the inverse of such a matrix is integral if and only if the determinant be ± 1. For orientable surfaces, one usually ignores the self-map that reverses orientation, and the mapping class group of the torus is then $SL(2,Z)$, the special linear group over the integers. For an orderly presentation of these matters, you should read Section 6.4 of John Stillwell's text [1980].

The mapping class group of the double torus is much more mysterious. For a thoroughly rigorous introduction you should read Joan Birman's monograph [1975]. There you will find the mapping class group of the double torus presented with five generators and two special relations in addition to those belonging to Artin's braid group. Let me list this presentation in a table. In the left column I have written the typical forms of the relators, without subscripts for easier typography. In the center are the corresponding verbal mnemonics. I shall explain the topological origin for the descriptive names *swap*, *coil* and *whorl* for the relators with pictures. In the right column uses the standard, subscripted notation.

Table 1. Presenting the Double Torus

$x, y, \ldots, z$	GENERATORS FOR THE DOUBLE TORUS	$x_1, x_2, \ldots, x_{n-1}$
$xz = zx$	REMOTE GENERATORS COMMUTE	$x_i x_k = x_k x_i$
$xyx = yxy$	ADJACENT GENERATORS SWAP	$x_i x_j x_i = x_j x_i x_j$
$(xy \ldots z)^6 = 1$	THE COIL HAS ORDER SIX	$(x_1 x_2 \cdots x_{n-1})^n = 1$
$xy \ldots zz \ldots yx = -1$	THE WHORL IS THE CENTRAL INVOLUTION	$w^2 = 1$ and $w x_i = x_i w$ where $w = x_1 x_2 \cdots x_{n-1} = x_{n-1} \cdots x_2 x_1$

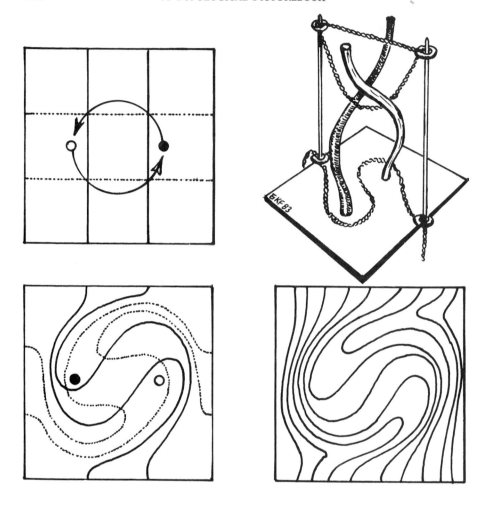

BRAIDING HOMEOMORPHISMS

Figure 2

Consider the strange device shown in 2(12). It consists of a square base with two rods rising up from it, gracefully bent so that each twists half way around the other. A string passes between the rods at the top. As the string drops to the base, its ends are constrained to move straight down. The bend in the rods guides the string so that it comes to rest on the base in an S-shape which winds half way around each rod. If the rods are removed from the base then there is an isotopy of the square which restores things as they were originally.

In 2(11) there are four strings crossing the square, two each way. The rods pierce the square at two punctures. As the square drops, the two punctures exchange places by moving along a circular track. This distorts the web as shown in 2(21). Note that it is only the black strings, passing between the punctures, which cannot be restored to their original position by pulling them tight.

Here is another way of imagining this self-map of the square. Draw a set of parallel lines across a square of elastic material which is fastened down along its perimeter. Press thumb and index finger over two points and twist half a turn. 2(22) shows the resulting distortion of the coordinate lines. This self-map of the square keeps the border pointwise fixed and permutes the punctures.

So self-maps of the punctured square are accurately described, up to isotopy, by the way *indicator paths* cross between the punctures before and after the mapping. This, in turn, is determined by which rod passes in front, and which in back. By convention, the twisting rods are called the *strands* of the *braid*, and labels $x, y, ..., z$ are attached to pairs of adjacent strands. A braid is coded by listing, from the top down, which way strands appear to cross: the left over the right, as in 2(12), has positive exponent, its inverse braid crosses the other way. Note that this sign does not depend on the way you look at the object in 2(12), because rods so twisted appear to cross upper left to lower right whichever way you see them. Historically, the convention is for the strands to twist like a left handed screw in order that the punctures rotate positively as they push the indicator curves about.

Artin Swap. *Figure 3.*

This composition of eleven diagrams portrays Artin's relator $xyx = yxy$, in three ways. The standard braid diagram is in the middle, 3(22). It is surrounded by indicator path diagrams for each of the two compositions of three self-maps. The start, 3(12), corresponds to the top level of the braids. The first half twist of xyx is coded by 3(11); that of yxy is coded by 3(13). After the third half twist, isotopies take 3(31) and 3(33) to 3(32). The corresponding "corkscrew" pictures, 3(41) and 3(42), combine these two ways of visualizing the relator. Note that the middle strand is free to move from one position to the other, taking xyx to yxy. That remote generators commute, $xz = zx$, is clear without pictures. Thus, Artin's braid group, B_n, has a presentation with $n - 1$ linearly ordered generators, wherein remote generators commute and adjacent generators swap.

Spherical Braids. *Figure 4.*

The cyclic permutation of the punctures, coded by the *product braid* $xy...z$, can be "undone" in two interesting ways. Compose this braid with its "flip", 4(11), and you have the *whorl* $xy...z^2...yx$. Here, the first strand winds once completely around the others. These may be bundled into a ribbon of

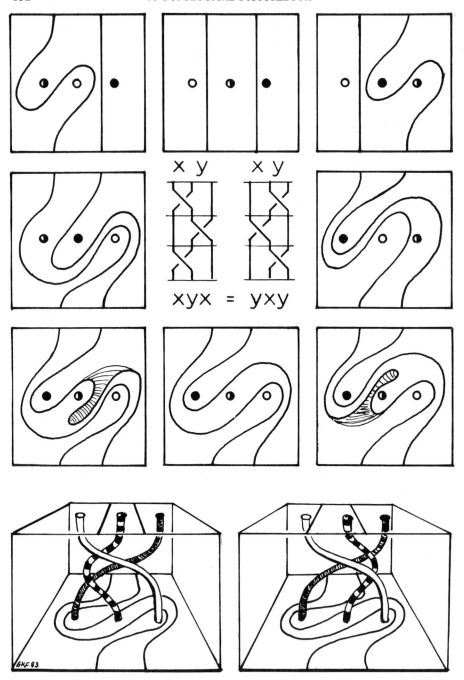

Figure 3   Artin Swap

CHAPTER 7 GROUP PICTURES 133

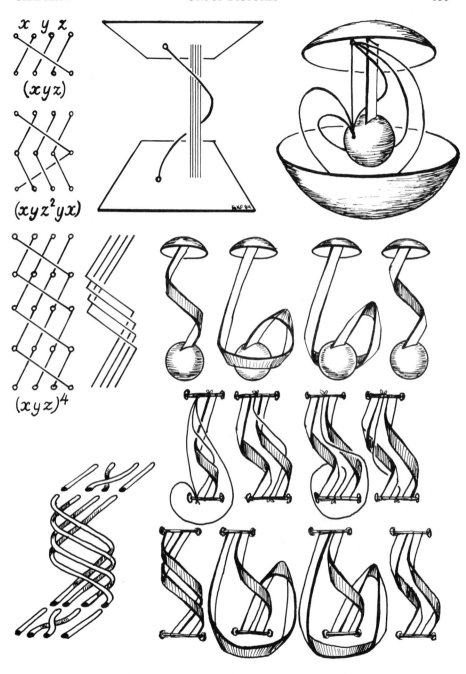

Figure 4 SPHERICAL BRAIDS

parallel strands which is not twisted or tangled, 4(12). Iterate the product braid, 4(21), until each strand returns to its initial location, and you have the *coil* $(xy...z)^n$. Here, each of the n strands winds once about every other. Together, 4(22), they bundle into a ribbon which makes one full twist in space. The coil generates the center of the braid group. The picture 4(31), of $y(xyz)^4y^{-1}$, shows how a half twist in two adjacent strands at the top of the coil can be combed down to cancel against its opposite at the bottom. Thus, it commutes with everything.

As long as the whorl refers to a self-map of the punctured plane, nothing can be done about its lone strand that twists around the ribbon. But suppose we interpret the braid as coding a self-map of the punctured sphere, 4(13). Then the odd strand can be untwisted from the ribbon by an isotopy of the space between two concentric spheres, which leaves the spheres themselves pointwise fixed. Thus, the whorl is a relator in the *spherical braid group* $B_n(S^2)$. In fact, it is the only new relator needed to pass from the plane to the sphere.

In the spherical braid group the coil has order two. As Magnus showed [1934], this follows both algebraically and topologically from the whorl. You can also see this directly by following the recoiling of the ribbon, starting at 4(23). As it lengthens, 4(24), the portion of the ribbon showing the dark side drops down, forming a second "collar" above the sphere. Turn both collars over, 4(25), and pull the ribbon tight again, 4(26). This twists it once the other way, representing the inverse of the coil. Consequently, a coiled ribbon with two full twists, uncoils without moving the ends relative to each other, while a single coil does not.

PLATE TRICK. *Figure 5.*

This remarkable property of space is splendidly illustrated by an old parlor trick with soup plates which I first learned from Ian Porteous in a Strasbourg restaurant. I have stylized the arm as a ribbon, to better illustrate the twisting. Start, 5(11), by holding up the plate, here with a two sided flag stuck in it to record the rotations. Proceed counter clockwise as shown, turning the plate twice counter clockwise in a horizontal plane. Half way, 5(42), your arm will have one full twist in it, which disappears, as you continue twisting your arm in the same way. Another version of this phenomenon was popularized by Dirac and explained by M.H.A. Newman [1942] with the help of braids. Dirac used it to illustrate how the Lie group $SO(3)$ of rotations in 3-space is doubly covered by the group S^3 of unit length quaternions. Note that the latter is the 3-sphere in 4-space.

Let me return to the whorl 4(13) for a moment, and show you how by moving one strand at a time you can transform the right-handed coil, 4(23), to its inverse, the left-handed one at 4(26). Three consecutive positions of the first strand are shown in 4(32). In 4(33), an intermediate strand has just

CHAPTER 7 GROUP PICTURES 135

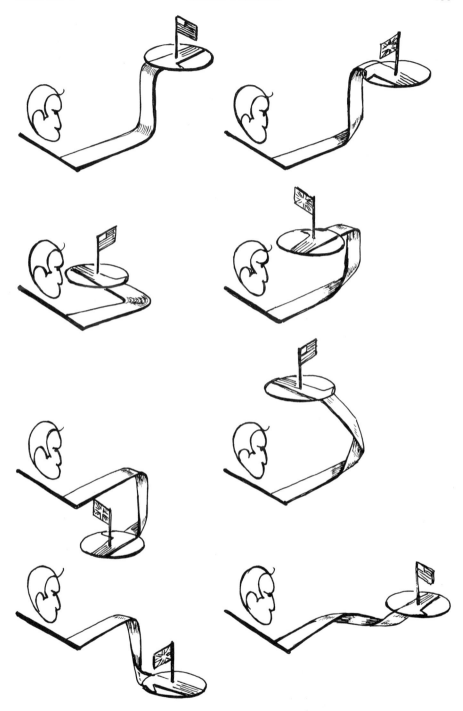

Figure 5 PLATE TRICK

been lifted off the right coil, comes around the bottom in 4(34), and is about to join its predecessors in the left coil.

In the mapping class group of the punctured sphere, however, the self-map coded by the coiled braid is already isotopic to the identity. The strands of the braid guide a coordinate web, as it falls from the celestial to the terrestial sphere, to isotopically equivalent positions, whether the earth is stationary or turned relative to the sky. In fact, as Fadell and Van Buskirk [1962] showed, this is the only new relation needed to pass from the spherical braid group $B_n(S^2)$ to the mapping class group of the punctured sphere, $M(0,n)$. Thus the former group is an extension of the latter, by a central element of order two.

WHORL AND CHIMNEY. *Figure 6.*

If you switch the roles played by whorl and coil above, and take the quotient of B_n by its center (the coiled braids) you obtain a different central extension of $M(0,n)$ by a cyclic center of order two. Magnus [1972] presents the natural history of this unnamed group. There it is labeled A_n^*. He traces its origin to Hurwitz' monodromy theory and describes how it acts as automorphisms of the fundamental group of the surface. That, in turn, defines a subgroup of the mapping class group of the surface $M(p,0)$, where $n = 2p + 2$. What follows is a sample of some pictures that can help in understanding the group theory. In particular, I want to show you what A_n^* has to do with whirling things about a spindle.

First consider Joan Birman's "favorite involution" of the p-holed torus T_p. There is a chalk drawing of it on the cover of this book. It consists of rotating the bilaterally symmetric version of this closed, oriented surface of genus p half a turn about the long axis. It has $n = 2p + 2$ fixed points, where the spindle pierces the surface. The orbit or identification space is the sphere. The n times punctured p-torus is a 2-fold covering of the punctured sphere. Over each of the singular points the covering projection, $T_p \to S^2$ is branched like $w = z^2$. In fact, the original way of imagining a Riemann surfaces was as a branched covering of the Riemann sphere.

There is an obvious set of $2p - 1$ invariant circles for the involution. They girdle the holes and tubes as they connect consecutive fixed points. Although the involution maps each of these *obvious circles* into itself, it reverses the orientation. On the sphere, the n branch points, arranged to lie on a meridian arc, are joined by segments covered by the obvious circles. You can pump up the layered sphere to form a singular image of T_p in space. The branch points become pinch points, and the *p + 1* odd numbered segments become the double lines joining the pinch points. In this context, these arcs are usually called *slits*. The even segments, call them *co-slits*, open up to simple closed curves on the (singular) surface. If you include the slightly less obvious invariant circle that goes all the way around T_p, it covers the meridian arc

Chapter 7 — Group Pictures

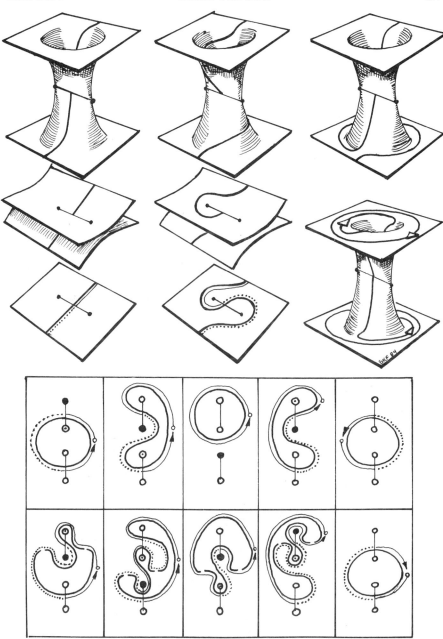

Figure 6 Whorl and Chimney

joining the poles of S^2 on the other side, then you can pump up the sphere the other way, switching the roles of slit and co-slit.

Consider the $n - 1$ self-maps obtained by twisting once about each of the obvious circles, as in 1(42). Later, I shall sketch just how mapping classes of surfaces in general can be generated by twists. For the double torus, $p = 2$, the five obvious twists already generate $M(2,0)$, which is isomorphic to Magnus' group A_5^*. That is why the mapping class group of the double torus has the beautiful presentation I put in the table above. Here is a way to see why the algebraic and the topological whorls are the same.

A *Dehn twist* of a surface is a self-map which is determined, up to isotopy, by events that occur in a neighborhood of a simple closed curve on the surface. Call it a *twisting circle* for the twist class. If, as is common, you depict this neighborhood as a right cylinder or a plane annulus, then the twist is defined as a cut-turn-glue operation which has a characteristic effect on a cross cut. Much insight is gained, however, if you imagine a 3-dimensional shadow of this neighborhood (located in 4-space) which looks like the *chimney* 6(11). The double line, where the chimney crosses over itself, is covered by the twisting circle. Flattening the chimney, 6(21), shows its relation to the slit of a branched covering of a disc 6(31). A diameter across the slit lifts to (one of two) cross cuts on the chimney.

An isotopy in space which keeps the top and base of the chimney fixed, but turns the pinched neck 180°, takes you to 6(12). Projecting the cross cut to 6(32) on the one hand, and letting it hang loose, 6(13), on the other shows that a half twist on the sphere corresponds to a full twist on the covering surface. This is what was meant in the footnote on page 165 of Birman [1974]. If you slide the twisting circle through the chimney, 6(23), from one sheet of the Riemann surface to the other, its projection on the Riemann sphere has its orientation reversed. In this picture, the central involution exchanges the two sheets by a vertical motion along parallels to the chimney axis.

Next, consider the plane diagrams which compute the effect of the succession $xy...zz...yx$ of twists associated with the whorl braid. In 6(41) you see a detail with four consecutive branch points, two slits and one obvious circle. You must imagine two copies of 6(11) placed over 6(31), and the two cross cuts joined into an indicator circle. A tour around the indicator, beginning at 3 o'clock is also marked. It begins on the upper sheet (solid mark), crosses the slit at noon, proceeds on the lower sheet (dotted mark) to the base of the later chimney, crosses its slit at 6 o'clock, and returns to the start. The indicator is isotopic to the (no longer so) obvious circle joining the middle two branch points. You should assume that twisting about this circle corresponds to one of the inbetween letters in the whorl, say b, in $xy...abc...zz...cba...yx$. Nothing happens to the indicator till the a-twist comes along and switches the black and grey dots, 6(51). An effective way of drawing such a diagram is to trace the previous curve except for a short piece crossing the slit. There, insert the switchback of the twist. Above 6(51)

CHAPTER 7 GROUP PICTURES 139

you should imagine one copy of 6(12) attached to one of 6(11). The isotopy 6(42) simplifies the indicator, which is now b-twisted to 6(52). Again an isotopy simplifies the indicator, 6(43), and shows that now it is immune to the effect of subsequent twists, $c...zz...c$. As shown, it lies entirely on the upper sheet of the chimney, but it is isotopic to the a-circle. As the b-twist comes around the second time, 6(53), it makes the indicator, 6(44), susceptible to the a-twist, 6(54), but to none of the remaining twists. The final 6(45) and initial 6(55) positions of the indicator coincide on the sphere, but not on the surface. As I have drawn it, the diagram encodes a surface detail with two chimneys, on which an isotopy takes the indicator in 6(55) to 6(45), reversing the orientation.

We know now that composing the whorl with the involution fixes the obvious circles pointwise. Since their complement on T_p consists in two topological discs, the composition is isotopic to the identity map on the surface. I leave to you the graphical exercise of computing the fate of the x-circle and the z-circle.

SWAPPING HANDLE CORES. *Figure 7.*

Mark a handle, 7(11), with two cross cuts, vertical and horizontal. Cut the tube and turn it as you would the sleeves of an old jacket, 7(21). Sew it back together, 7(31), and thread a loop through the handle. Now pull the handle to the front, 7(32), and give it a quarter turn, without moving the square border, 7(22). This restores the initial position. A self-map of the handle is defined by a superposition of 7(12) onto 7(11). As the shape of the cross cuts indicate, the mapping class defined by this procedure is not the identity.

SWAP DIAGRAM. *Figure 8.*

The same procedure is diagrammed abstractly in the top row of the next figure. Detail 8(11) corresponds to 7(11), 8(12) to 7(21), 8(13) to 7(32), and 8(14) to 7(12). In 8(15), I have separated the handle from an annular neighborhood of the border, called its *collar*. Iterate the swap four times. 8(21)-8(25) shows that the handle itself returns to the original position, while 8(31)-8(35) shows that a twist has been visited on the collar. Thus the 4th-power of a swap equals a twist about a circle remote from the handle. The last row shows how the swap is the product xyx of twists about the generating circles, 8(41), of the handle. In 8(42) the horizontal indicator has been twisted about the vertical generator. An x-twist produces 8(43). An isotopy, which you should think of as pulling the black headed indicator path taut, slides it through the tunnel and a little way up the other side, 8(44). A y-twist, its support is dotted in, produces 8(45). This, in turn, is isotopic to

140 A Topological Picturebook

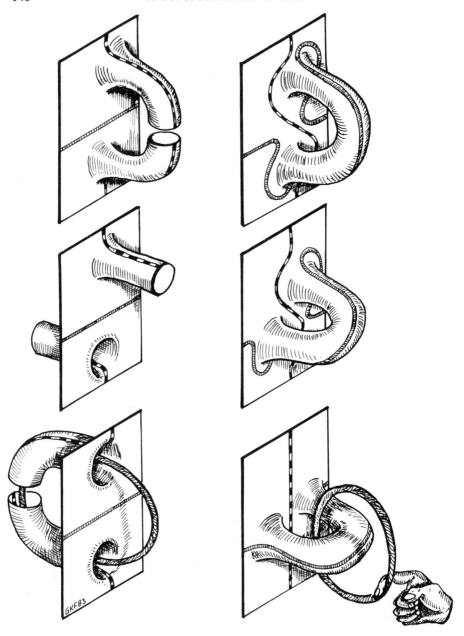

Figure 7 Swapping Handle Cores

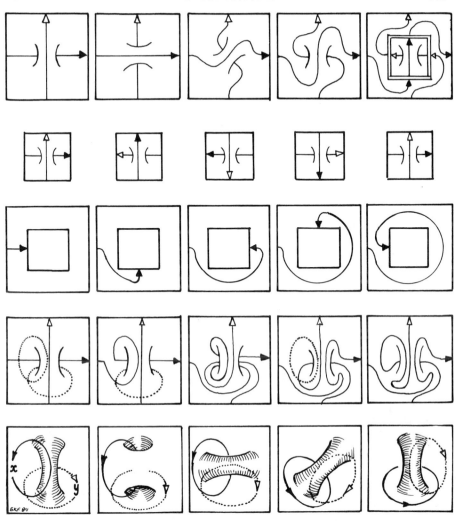

Figure 8 SWAP DIAGRAM

8(14) by shrinking the indicators. Since both self-maps of the handle do the same thing to the cross cuts they belong to the same isotopy classes.

The last row shows that the handle generators swap this way

$$(x, y) \to (-y, x)$$

where $-y$ denotes the circle with its orientation reversed. 8(51) corresponds to 7(11), 8(52) to 7(31), 8(53) to 7(32), 8(54) to 7(22) and 8(55) to 7(12). After an isotopy on 8(55) which pulls the white tipped circle through the tunnel to the left side, a superposition of 8(55) on top of 8(51) defines the self map of the handle.

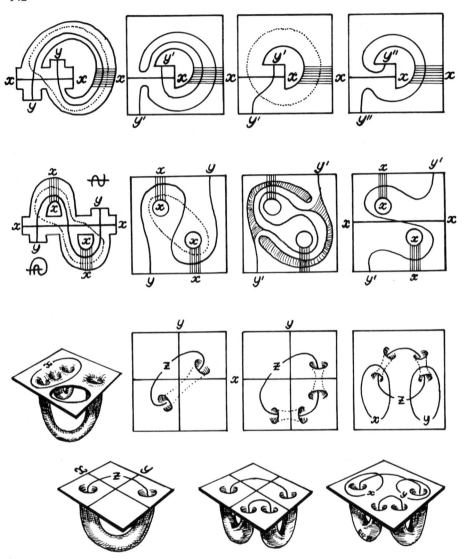

LICKORISH TWISTS

Figure 9

Swapping the core circles of handles on a compact surface F plays a major role in demonstrating that the mapping class group of F is generated by a fixed, finite set of twists. This theorem was discovered independently by Max Dehn [1938] and W. Lickorish [1964]. Lickorish included non-orientable (= one-sided) surfaces in his investigations and used the term *homeotopy group* for the mapping class group. For the purpose of proving the theorem, he invented a number of very clever topological constructions. They are part of the more compact proof starting on Page 168 of Birman [1975]. To illustrate these ingenious moves it suffices to prove only the weaker, first part of the theorem. This says that every self-map of F is isotopic to *some* product of twists without fixing a particular basis which generates every self-map.

Suppose $y = h(x)$ is a self-map of F which moves a circle x on top of a circle y. Suppose we show further that a finite composition h', of twists returns y to x. Then x is fixed by the composition $h'h$ and its mapping class is associated with a particular self-map of the bordered surface F_x obtained by slitting F open along x. If F_x is connected, then it is truly simpler than F, and inductive arguments, starting from a disc with holes (alias punctured sphere) can be invented. It is also interesting that the condition that x should not separate F enters into the discovery of h' itself. Regarding the orientability of F, what is really at issue is whether a given circle is one- or two-sided, i.e. whether a narrow ribbon neighborhood of it on F is a Möbius band or not.

Finally, if you are new to these matters, remember that twists about isotopic circles are in the same mapping class. This allows me to shape the pictures conveniently. In describing the effect of a homeomorphism on a surface, it is a good idea to imagine a transparent, flexible copy of the surface lying on top of the fixed copy. The target circle x is on the fixed copy, the working circle y is on the flexible copy. A small isotopy will move y into general position with respect to x, so that y crosses x from one side to the other at a finite number of points. There are global isotopies that reduce such crossings to a minimum, even among an entire generating set of circles, but for present purposes we do not need this.

Suppose first, that x is 2-sided and y crosses x only once. Then the pair generate a handle on F and a swap takes y to x. The difficult cases are when x and y are disjoint, or when y crosses x twice but from opposite sides of x. Think of two circles in the plane. Otherwise y crosses x many times and here is a way of twisting y to y' so that the new circle crosses x fewer times.

For the second step, suppose y crosses x from the same side at two consecutive points. That is, x has an arc x^* which is crossed by y only at its endpoints, though either of the corresponding arcs, y^*, of y can cross x elsewhere, 9(11). The effect y' on y produced by the twist on the ribbon along the circle $x^* + y^*$ looks like 9(12). The isotopy 9(13) reduces the number of times y' crosses x by at least one, and it also removes crossings far away from the arc x^*. It is a little more elegant to twist again, 9(14), and thus remove the crossings at the ends of x^*. Elsewere, the crossings are the same, though the order on the image of y^* has been reversed.

This way, consecutive y-crossings from the same side of x can be twisted out of y pairwise, until y switches sides at each crossing of x. Suppose, thirdly, there are (at least) three of these, occurring at the ends and midpoint of an arc x^* of x, 9(21). Starting at one endpoint of x^*, the arc y^* of y wends its way two more time across x^*. In the case shown, x^* and y^* end at the same place. I leave the other case to you. Once the appropriate twisting circle (dotted) is identified in 9(21), it is convenient to restyle the neighborhood of the circle $x^* + y^*$ like 9(22) in order to compute the effect y' produced by the twist on y, 9(23). An isotopy takes this to 9(24), which shows that y' differs from y only in the neighborhood of $x^* + y^*$, and that the image of y^* no longer crosses x^* at the endpoints.

This brings us to the last two cases, that y crosses x either twice but from opposite sides of x, or not at all, and that nothing can be done about this by an isotopy on F. Here is where the property of not separating F comes in. Build yourself the following model of the surface. The plane of the paper represents the sphere, endowed with a certain number of handles and holes. These are represented by punching out little discs here and there, some of which are pairwise connected by underground tunnels, 9(31). Suppose x and y cross twice. Then represent them by the axes crossing at the origin and inifinity. Since neither x nor its homeomorph y may separate F, both vertical and both horizontal half planes must be connected by underpasses. If this occurs economically, with a single tunnel as in 9(32), a helper circle z can be conducted through the tunnel, and then on the ground to cross x once and y once only. Now swap y to z and z to x, recalling that swaps are twist compositions. If two tunnels are needed, 9(33), the helper circle comes up for air, but otherwise does the trick the same way. If x and y do not cross, draw them as circles in the plane with disjoint insides, before marking the manholes. Since some tunnel connects the inside of x to its outside, and some tunnel connects the outside of y to its inside, if it isn't the same tunnel, it must look like 9(34). In either case, there is a helper circle across which to swap y over to x.

KING SOLOMON SEAL. *Figure 10.*

For the general and very topological method of computing mapping class groups, $M(p,q)$, of surfaces with p handles and q holes, consult Hatcher-Thurston [1980]. It was the preliminary, even more topological version of their work, as reported by A. Marin [1977], which inspired the graphical tricks I have used in this chapter. See Wajnryb [1982] for an explicit, algebraic presentations for $M(p,0)$ and $M(p,1)$. As current research in this area becomes more algebraic, pictures become less welcome, except as pedagogical aids. So, let me close my story about pictures from group theory by looking once more at the mapping class group, $M(1,0)$, of the simple torus, its representation as $SL(2,Z)$ and some of its more geometrical presentations as an abstract, two generator group.

If you choose generators of a handle as in 8(51) then there are two ways of merging the circles by breaking their crossing node. Apropriately (re)oriented, each is the image of one generator twisted about the other. The twist of acircle about itself is, of course, isotopic to the identity. So, the following table summarizes the different ways of thinking about the same thing.

$$\left.\begin{array}{l} x = x^x \\ x+y = y^x \end{array}\right\} \Rightarrow x = \begin{vmatrix} 1 & 0 \\ 1 & 1 \end{vmatrix}$$

$$\left.\begin{array}{l} x-y = x^y \\ x = y^y \end{array}\right\} \Rightarrow y = \begin{vmatrix} 1 & -1 \\ 0 & 1 \end{vmatrix}$$

For conceptual simplicity and typographical economy, I shall use the same letter for the generator of the mapping class group, the corresponding twist and its twisting circle. The additive notation in the first column refers to the operation in the free abelian fundamental group of the torus. Twisting y about x, i.e. the result of applying the x-twist to the y-circle, is written like this y^x in the second column. I prefer the power notation to indicate the image under a function because it preserves the natural order of composition. In the last column, composition in $M(1,0)$ becomes multiplication of matrices in $SL(2,Z)$, and in the same order.

Let me show you how this algebra works by applying it to the self-map $r = xy$, which I call the *roll* because it expresses a six-fold rotational symmetry of the torus. In checking the following topological computation only the effect on the $x + y$-circle of the y-twist should require pencil and paper.

$$x^{xy} = x^y = x - y$$
$$y^{xy} = (x+y)^y = x^y + y^y = x - y + y = x$$

This verifies the matrix product for the roll, which is the first entry in the following table of interesting identities. The swap matrix can be factored two ways, confirming Artin's relation. In the case of the torus, the third obvious circle invariant under the canonical involution happens to be isotopic to first. The isotopy $z = x$, Artin's relation, the associative rule from group theory and matrix algebra reveals the whorl to be at once the square of the swap, the cube of the roll and a square-root of the coil.

$$\text{ROLL} \quad \begin{vmatrix} 1 & -1 \\ 1 & 0 \end{vmatrix} = \begin{vmatrix} 1 & 0 \\ 1 & 1 \end{vmatrix} \cdot \begin{vmatrix} 1 & -1 \\ 0 & 1 \end{vmatrix} = xy = r$$

$$\text{SWAP} \quad \begin{vmatrix} 0 & -1 \\ 1 & 0 \end{vmatrix} = \begin{vmatrix} 1 & -1 \\ 1 & 0 \end{vmatrix} \cdot \begin{vmatrix} 1 & 0 \\ 1 & 1 \end{vmatrix} = rx = xyx = yxy = s$$

$$\text{WHORL} \quad \begin{vmatrix} 0 & -1 \\ 1 & 0 \end{vmatrix} = xyz \cdot zyx = \begin{cases} xyx \cdot yxy = s^2 \\ xyx \cdot yxy = xy \cdot xy \cdot xy = r^3 \end{cases}$$

$$\text{COIL} \quad \begin{vmatrix} 1 & 0 \\ 0 & 1 \end{vmatrix} = (xyz)^4 = \begin{cases} xyx \cdot xyx \cdot xyx \cdot xyx = s^4 \\ xyx \cdot yxy \cdot xyx \cdot yxy \end{cases}$$

$$= xy \cdot xy \cdot xy \cdot xy \cdot xy \cdot xy = r^6$$

$$= (xyx \cdot y)^3 = (st)^3$$

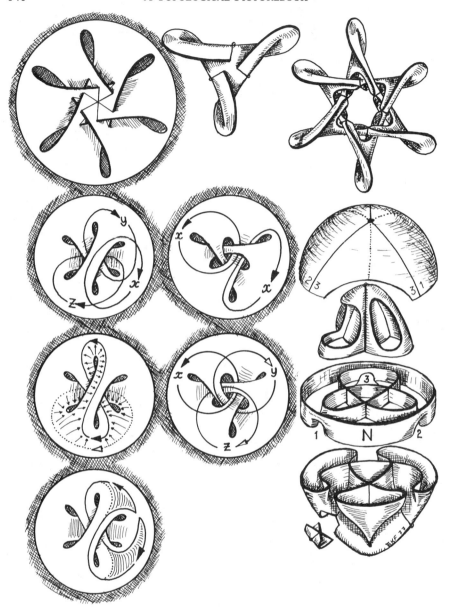

Figure 10 KING SOLOMON SEAL

In the last line of the table I have used the letter t to stand for the twist y, though x would do just as well. This table explains the topological connection between the following popular presentations of $SL(2,Z)$:

$$\langle s,t \mid s^4 = s^2ts^2t^{-1} = (st)^3 = 1 \rangle$$
$$\langle s,r \mid s^2 = r^3 = -1 \rangle$$
$$\langle x,y \mid xyx = yxy, (xy)^3 = -1 \rangle$$

A twist about a circle that separates a surface is isotopic to the identity. You can convince yourself of this by representing the surface as a sphere with holes and handles (even cross caps) and twisting along the equator. Smoothly turning the northern component once clockwise against the southern component undoes the cut-turn-paste operation of the twist. Recall how Lickorish moved any (2-sided) nonseparating circle onto another by a composition of twists. Hence the twist maps are conjugate to each other. Therefore, on the torus, they must have the form:

$$\begin{vmatrix} a & b \\ c & d \end{vmatrix}^{-1} \cdot \begin{vmatrix} 1 & 0 \\ 1 & 1 \end{vmatrix} \cdot \begin{vmatrix} a & b \\ c & d \end{vmatrix} = \begin{vmatrix} 1-ab & -b^2 \\ a^2 & 1+ab \end{vmatrix}$$

In particular, a self-map which is a twist about some circle, has trace 2. The above table thus shows that neither swap, whorl nor roll are twist maps.

You have already met the toroidal symmetries of order 2 and 4 associated with the whorl at the beginning of the chapter. Here is a way of visualizing also the 6-fold symmetry associated with the roll, and the 3-fold symmetry of its square, which occupies such a prominent place in Hatcher-Thurston [1980]. The picture is based on the excellent map of the torus to the sphere whose image folds along two transverse deltoids arranged in the shape of *Solomon's Seal*, see Page 191 of Francis-Troyer [1977].

Pump up this image of T^2 on S^2 as follows. Link the two singular ribbons as in 10(13). Each has three pinch points, a deltoid contour and one window. Recall that his means that the border spans an embedded disc which is fully visible. In 10(12) this window has been closed. The other two borders of 10(13) fit into the surface 10(11). Let me deform this into an embedded annulus with three bridges 10(21), which has Euler characteristic equal to −3. Hence the surface F, obtained by sewing three discs into 10(11), is indeed a (singular) torus in space, and it has 6-fold rotational symmetry. The clockwise turn by 60° of F can also be visualized on 10(21) if, after the rotation, you move the bridges vertically by a regular homotopy. The self-map permutes the three obvious circles, x,y,z, taking y to x and x to z. An isotopy moves the z-circle into the figure-8, 10(31), by moving it across one of the discs, like 10(12). The figure-8 is isotopic to the circle obtained by breaking the crossing node of x and y, 10(11). This self-map is, therefore, the roll, $(x, y) \to (y, x - y)$.

Novel in Hatcher and Thurston's original approach to computing $M(p,0)$, as related by Marin [1977], was their use of generic deformations of Morse functions on T_p to discover the relators. For genus p, there are p circles which, when cut open, reduce the surface to a sphere with $2p$ holes. Fix such a *cut system*, C. The subroup H of $M(p,0)$ which permutes the circles of C can be computed in terms of extensions of $M(0,2p)$, the mapping class group of the punctured sphere. One additional self-map, a swap, is needed to generate $M(p,0)$ from H.

Consider the height function on the various positions of the handle in Figure 7. The indicated cut circle of the handle remains horizontal, up to an isotopy, until 7(22). In 7(32) it runs around the inside of the handle, and cannot remain horizontal as the handle is turned 90° to 7(12). In 7(22), the two critical points exchange heights. This is seen better in 8(54), the horizontal cusps of the astroid are the two saddles.

Thus, during a deformation of the height function, whenever a cut circle cannot remain isotopically level, it is swapped with its fellow generator of the handle. So generic paths of Morse functions translate into elements of the mapping class group. A very important relator derives from the situation that the function has, momentarily, three saddles at the same height. I have drawn the regular neighborhood N of the level as an immersed annulus with three bridges, 10(33), and 3-fold symmetry. You should imagine a hemisphere, 10(23), attached to N along the circular rim of N. Height functions near the given one are faithfully parametrized by their maxima near motion north pole, as the surface is made to wobble a bit. A 1-parameter family of functions is recorded by a path on the polar cap. As the function "crosses" one of the three solid radii, two of the saddles are at the same height. The same happens at the dotted radii, but being "inessential," no choice between competing cut circles is necessary.

Note, in passing, that 3-saddled N has all three borders horizontal, hence they could be capped by three extrema of the height function. Thus, by Möbius-Morse theory, the surface has genus one; the closed surface is a torus. The cone on the trefoil, 10(23), is an equivariant Whitney umbrella. It may be considered a shadow of the graph of $w^2 = z^3$. The lower cap, 10(43), with three pinchpoints, unfolds this singularity.

Since N has genus one, any one of the three obvious circles, x,y,z, passing over two of the saddles could be a member of a complete cut system, C, on the general surface, F, 10(32). To see that a 120° turn of N is equivalent to the square of the roll, first move the x-circle across the norther hemisphere, 10(22). Then observe that x goes to $-y$ and y goes to $z = x - y$, as predicted.

There is, of course, much more to tell and draw in group theory, and it cannot be done all at once. That the basic relators of the mapping class group reappear in the least likely places is illustrated by my next, and final, picture story.

8
THE FIGURE EIGHT KNOT

My final picture story is about visualizing how a knot complement fibers over the circle. This exercise of the imagination consists of filling the void of space, closed up by a single point at infinity to form the 3-sphere, with a continuous succession of surfaces spanning the knot. That is, through each point not on the curve, there will pass a unique copy of a surface whose boundary is the knot.

The reason for wanting to do this comes from the project of imposing a non-Euclidean (hyperbolic) geometry on most 3-dimensional manifolds, as explained in the survey articles by Bill Thurston [1982] and John Milnor [1982]. The complement of the figure-8 knot plays a major role in Thurston's unpublished textbook. For my story about this knot, I used two manuscript editions [1977,1982] of the text. It is important to realize that the examples in this chapter belong to an area of topology that already has a highly developed and effective graphical shorthand. This style consists of "schematic" diagrams which are quite terse and frequently incomprehensible to the novice. Recognizable pictures of familiar shapes, artistically arranged in space as dictated by the abstract diagrams, are sometimes welcome. That is the purpose of my efforts here.

In Thurston's grand scheme, it is the topology of a manifold that limits and frequently determines its possible geometries. For simplicity's sake, the text that goes with my pictures is topological. Here is a brief note about the group theory and geometry which I have left out. The way to get a geometrical manifold is to take a polyhedral chunk from a geometrical space and identify its faces pairwise with each other. In order that the geometry matches up properly across the faces and around the edges, it is best to let a group of isometries on the ambient space dictate the identification. In the story about the Penrose tribar the polytope was a cube in Euclidean space, and the isometries were rotations and translations. In 1912, Gieseking identified the faces of an ideal tetrahedron in hyperbolic space as dictated by a group of orientation reversing isometries, see Magnus [1974, p.153ff]. By finding a suitable representation of the fundamental group of the figure-8 knot complement, Riley [1975] found a hyperbolic manifold homeomorphic to the knot complement. Troels Jørgensen [1977] used the way the knot complement fibers over the cirlce to demonstrate its hyperbolic structure. Thurston showed how it is also the (orientable) double cover of Gieseking's manifold.

The way from the knot complement to the gluing diagram on a hexahedron and the way back from a diagram to the complement is not part of the fibering story. I have included it here because it is such a rewarding subject for descriptive topology. This is also true of the digression on the Hopf fibration at the end. Hopf's decomposition of the 3-sphere into a bundle of 1-dimensional fibers (circles that link) that arrange themselves exactly like the points on a 2-sphere lead to Seifert's [1933] theory of (singular) fibrations of 3-manifolds over surfaces. The theory provides a complete classification of the 3-manifolds with the geometry of the 3-sphere.

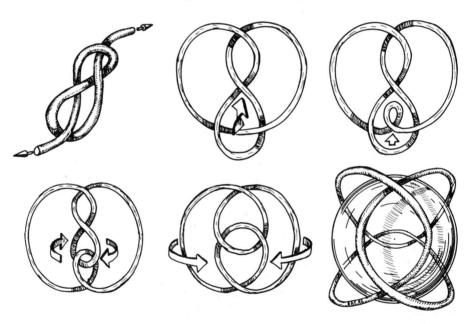

Projections of the Knot

Figure 1

As Tait observed [1911], the figure-8 knot, by sailors "used only to prevent ropes from unreeving; it forms a large knob" has been part of "topology" ever since Listing [1847] coined this name for the mathematical theory of position. Tait called it the "four-knot" because it is the only simple closed

CHAPTER 8   THE FIGURE EIGHT KNOT   151

knotted curve in space whose picture can be drawn with four alternating over and under crossings. Here is a recipe for moving from one of its planar positions 1(12) to the other 1(22). Twist a small loop into the segment of the knot which passes through the lower loop of the "eight" in 1(12) to obtain 1(13). Shrink the larger loop to link the smaller, 1(21), and untwist it, 1(22). To place the knot into its most symmetric position in space, 1(23), swing the outside arcs forward. A rotation by 90° in the picture plane, followed by a reflection in that plane, moves the knot 1(23) into itself: it is its own "mirror image." Invariance of a geometrical objects under such a *turn-reflection* (Drehspiegelung) is quite useful. Here it will eventually halve our effort in visualizing the fibration of its complement by Seifert surfaces.

HEXAHEDRAL COMPLEMENT. *Figure 2.*

It is interesting to see how the complement of the knot in the 3-sphere may be regarded as the result of gluing the faces of two tetrahedra in a particular way. This example is based on the procedure in Chapter 1 of Thurston's [1978] text , but I shall apply it to Seifert's [1934] classic position 1(11) of the knot rather than the symmetric position 1(23).

Imagine dipping the knot into a viscous liquid which adheres to it, forming the membrane 2(11). This membrane coincides with the viewing plane outside the vicinity of the knot. On the knot, the membrane may be reconstructed as follows. Span the knot by a twisted disc which bites itself along a horizontal double-segment, 2(12). In this detail, the immersed disc is severed by a vertical diameter at 6 o'clock to form two embedded discs whose borders link. The knot is also spanned by another singular disc which passes through infinity in S^3. It is the turn-reflection of the first disc, if looked at in the symmetric position, 1(23). In 2(13) I have severed this immersed disc by a horizontal diameter at 12 o'clock. These two discs fit together, as suggested by 2(22), to form the membrane, with a horizontal and vertical seam collecting six sheets of the surface. 2(21) shows the horizontal seam. The seams are oriented to the right, respectively upward, and coded with a white, respectively black, arrowhead. A cross sections of the membrane by a plane transverse to the seam looks like three concurrent lines. Thus, there are three pairs of opposite dihedral angles at each seam, 2(23).

Now split the seams to produce the four embedded, five-edged discs 2(22). Two adjacent edges form a slit, these are flanked by segments of the knot, and the fifth edge is on the other seam from the one forming the slit. You see *W,E,S*, the western, eastern and southern discs respectively, in their entirety. The northern one, *N*, which goes through infinity, is represesented by two pieces, one near each seam. Since each disc has a half-twist (two are right handed, two left) a portion of both sides of each penta-lateral faces you, the observer.

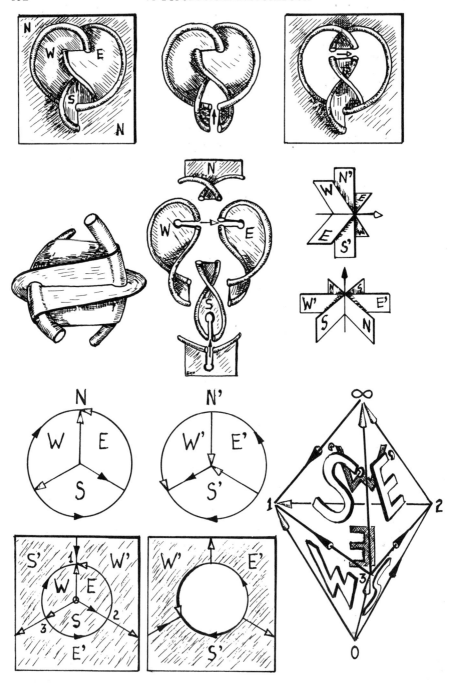

Figure 2 HEXAHEDRAL COMPLEMENT

CHAPTER 8   THE FIGURE EIGHT KNOT   153

Use the dihedral angles in 2(11) to read the front tetrahedron 2(31). The fully visible horizontal dihedral reads *WE* (not *EW*, because the white arrow points to the right). Look at 2(23). The dihedrals *SW* and *ES*, so oriented, can be read from these white-, respectively black-arrowed paddle wheels, 2(23). The other tetrahedron corresponds to the half-space on the far side of the membrane. The backside of the membrane, labeled with primed letters, yields the reading 2(32) for the other tetrahedron. To fit the two polyhedra together along their northern faces, invert the back tetrahedral gluing data, 2(32), to form 2(42). The northern face has been omitted, leaving a circular hole into which to fit the front tetrahedron 2(31). This yields the abstract gluing diagram 2(41), which translates into the hexahedron on the right.

HEXAHEDRAL GLUING DIAGRAM. *Figure 3.*

The passage from knot complement to polyhedral gluing diagram is reversible. In Chapter 3 of the 1982 edition of Thurston's text there is an example of a hexahedral gluing diagram, different from the one above, which produces the complement of a knotted surface of genus 2 with 3-fold symmetry: Thurston's "tripus." Here I shall use Thurston's procedure to recover the figure-8 knot complement from the gluing diagram.

Diagram 3(11) shows 2(41), but with edges widened into strips, and vertices expanded into poly-lateral patches. You should imagine you are looking down on the top of a half-space. This half-space is a solid ball which passes through infinity. Eventually, the three pairs of faces must be glued together, and with the orientation indicated by the black and white dots. The black dots correspond to a "sink" and the white dots correspond to a "source" for the arrows on the hexahedron 2(43). Now, the gluing diagram 2(41) also specifies the two sets of edges on 2(43) to be identified. They are marked with 5 white arrowheads and 4 black ones.

This requirement translates to 3(11) as follows. The strips (striped rectangles) are to be glued to fluted, cylindrical rods representing thickened line segments. The white rod attaches to five strips; the black rod, only four. Now we come to a fundamental change in one's point of view: a cylinder is also a thick disc. The two rods flatten into coins and the strips change accordingly, 3(12). Next, 3(13), the faces shrink to hexa-lateral *tees*. The tees are connected by cross-hatched *roads*, which are destined to be attached to the coins. Two pairs of tees, *WW'* and *EE'*, are properly oriented (dotted lines) so that they can be glued together. Bring them into proximity, 3(21), by turning the outer rim 120° counter-clockwise.

The center detail, 3(22), shows how the roads are reconnected after the eastern tees have been identified. Note how the gluing handle extending into "our" space (top) can be pushed down, forming a worm hole extending into "their" space (bottom).

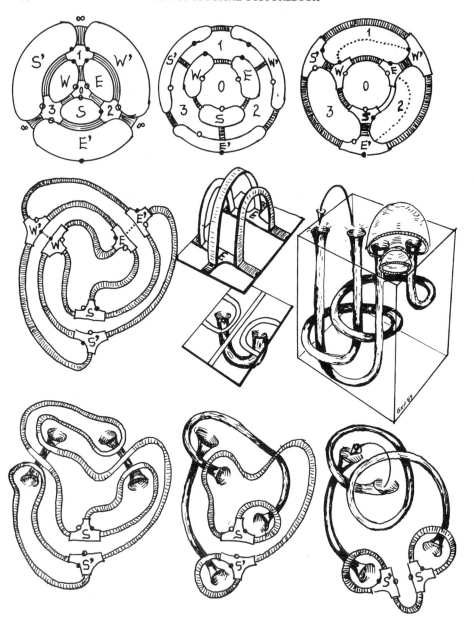

Figure 3   Hexahedral Gluing Diagram

Globally, 3(21) becomes 3(31). Unlike the western and eastern tees, the southern tees are oriented the wrong way for gluing. If, nevertheless, you were to glue them together in the position shown, and attached the coins, you would have Thurston's "tripus." To prepare the southern tees for gluing, deform the roads, dragging the subterranean tubes along by their vents, first to 3(32), and then to 3(33). In the last move, the road loop at the tee marked S shrinks along itself, dragging the tunnel over itself to produce two more over-crossings. Now that the southern tees are properly aligned in 3(33), you can draw the final road map for yourself.

Now bring the entire solid into finite space. 3(23) shows the block, with the plumbing visible from the side. Two bubble shaped coins have been affixed to the two circular roads remaining after the handle was attached to 8(33). The larger bubble, the one with a dent, insures that the tubes join into a single knot. I leave to you the graphics to check that 3(23) is indeed the figure-8 knot complement.

SEIFERT SPANNING SURFACE. *Figure 4.*

It can be quite difficult to see how a surface might span a knot which is drawn in a particular way. For this reason it is wise to experiment with several positions of the knot. Position 1(12) is not the most convenient but serves to illustrate a simple approach to spanning a knot. There is a systematic procedure for doing this, due to H. Seifert [1934], based on the combinatorics of the knot projection, but this won't be needed here. With a bit of practice you can learn to "see" twisting ribbons, as well as more or less convex patches, spanning parts of the knot. The difficulty is to see how these patches might fit together. In 4(11) two ribbons, with a full twist each, have come into conflict. They attach to "opposite" sides of the knot. Two ways of resolving this conflict locally are shown in 4(22). 4(21) is a line pattern for this picture. The conflict is resolved globally in 4(12) by continuing the lower ribbon so it fills out a bowl-like patch. This forces the upper ribbon to become a handle or strap across the bowl; the result is the classic position as it appears on page 584 of Seifert [1934]. For a different version, 4(13), continue the upper ribbon of 4(11) to form a plane patch extending to infinity beyond the square border. This forces the lower strap to hang limply from the patch.

The twisted straps of 4(11) could have been merged differently so that the bowl of 4(12) is in front of the upper strap, and the square patch of 4(13) is in front of the lower strap. This specifies four surfaces spanning the knot. These surfaces can be described in terms of 4(23), which is a "cubist" stylization of 4(12). For 4(13) draw the box in front. For the alternate version of 4(12) draw the horizontal strap inside the box; draw the box in front for the alternate of 4(13). The isotopy in S^3 which takes 4(13) to 4(12) is easy to imagine. Move the infinite, planar patch of 4(13) past infinity. Think of

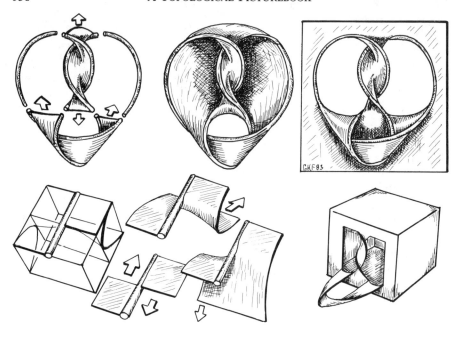

Figure 4 SEIFERT SPANNING SURFACE

a huge bowl behind the plane of the paper with a square opening into which 4(13) fits. Shrink it to the shallow bowl of 4(12). Now, if you can already imagine an isotopy of 4(12) into a position where the bridge is in back of the bowl, and a similar isotopy through the fourth position back to 4(13), then you have already anticipated the end of this story. Remember, the motion must not involve any tears or self penetration of the surface, and each surface must differ from every other in all of its interior points. The boundary is, of course, always the knot K. Moreover, Seifert's surface turns about K without sliding along it: a page in a ring binder, not a spiral notebook.

SIX HAKEN SURFACES. *Figure 5.*

An incompressible and boundary incompressible surface F in a 3-manifold M with boundary ∂M is one that cannot be simplified by closing off a tube, as a torus in 3-space is simplified to a sphere by gluing in a meridian disc (or longitudinal disc through the hole in the donut). In recognition of Wolfgang Haken's important contributions to manifold theory, in which such surfaces play a crucial role, their tongue twisting-name should be shortened to *Haken surfaces*. The technical definition is that the boundary inclusion map $\partial F \hookrightarrow \partial M$, as well as the interior inclusion $(F-\partial F) \hookrightarrow (M-\partial M)$, both induce monomorphisms on the respective fundamental groups. To the working topologist, this means that a loop on F which can be spanned by a disc in M, already bounds a disc on F itself.

In Chapter 4 of the 1977 edition of his text, Thurston shows that there are exactly six distinct Haken surfaces in the complement of the figure-8 knot. Here is a way of generating them from our initial knot-spanning membrane. It involves a *seam splitting* procedure I shall describe first. Take a wire segment and bend it nearly into a plane circle, but then bend the ends out so that they could be continued as parallel tracks. Span this lyre-shaped hook with a disc which has a free border segment and which could be continued as a ribbon between the tracks. Link two of these, 5(11), one horizontal and one vertical, so that the two free border segments are perpendicular to an axis running parallel to and between the continuing tracks. The two discs cross each other along a double segment on the axis.

This double segment may be split lengthwise in two ways to yield an embedded annulus spanning the linked lyres. In 5(12) the split goes through the visible dihedral along the seam. The surface shrinks to 5(13), then to 5(21), which can be drawn without cusps and has the original symmetry. That is, the figure is invariant under a turn-reflection in the plane orthogonal to the axis. Note that the ribbon through the free ends splits temporarily into two, each twisting to the left.

If you split the seam the other way, lengthwise through the invisible dihedral, then by symmetry the two ribbons must twist to the right. The trick is to draw the surface with the original wires in place, 5(22). Think of this surface as obtained by continuing the ribbon through the hook with a quarter right twist, and then bridging opposite shores by a small ribbon with a full (right) twist, as in 5(23).

Now span the figure-8 knot by a disc that bites itself, 2(12), and split the seam on the visible dihedral, 5(31). Once you "straighten" this surface out a bit, 5(32), you will notice the two Möbius bands on it which are characteristic of a punctured Klein bottle. Splitting the seam the other way, 5(33), produces the punctured torus which you have already studied in 4(12). If we split the seam of 2(13) in the two ways, we again obtain the non-orientable, 5(41), and the orientable, 5(42), surfaces of genus one. Compare the latter with 4(13) and you will agree that the two "holey" tori are the same (isotopic). But the two "holey" Klein bottles are not.

To see that surfaces 5(31) and 5(42) are not isotopic in S^3, compare the twisting of the curve along which it is attached to the (thickened) knot. You can easily check the twist sense by observing the way a marker curve, running along the visible front of the knot tube, pierces the surface. In the detail, 5(43), of the northern most over crossing in 5(31), the marker is directed the way right-handed people make a figure-8. The surface twists to the right. The corresponding detail of the same crossing for 5(42) twists the other way (draw it!). Check that this holds at each over-crossing. The putative isotopy between the two surfaces would, of course, preserve the sense of this twisting.

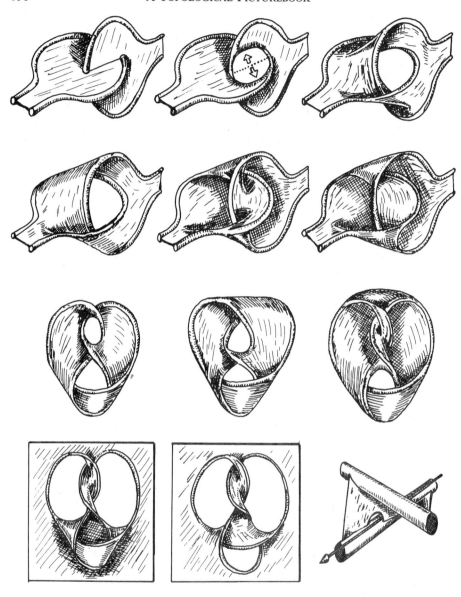

Figure 5 SIX HAKEN SURFACES

Here, then, are the six Haken surfaces. The toroidal skin of the knot tube is the sole closed Haken suface. Thicken each of the holey Klein bottles and peel off the skin, to produce two holey tori. These differ from the Seifert surface because they go around the knot twice. The Seifert surface is the sixth and I shall continue with its story.

Chapter 8 — The Figure Eight Knot

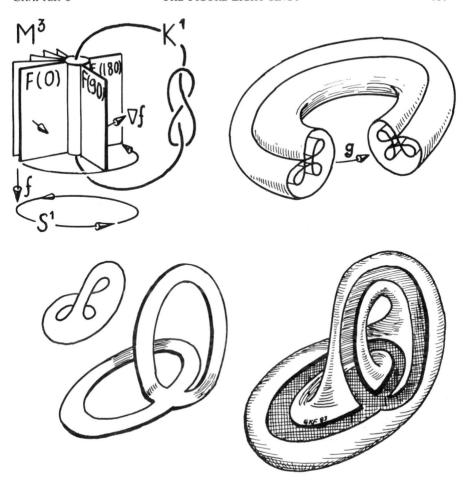

Fiber Mnemonics

Figure 6

A fibration "over the circle, by Seifert surfaces" of the knot complement $M = S^3 - K$ may be described by a smooth map $f:M \to S^1$ of M onto the unit circle, which is everywhere of maximal rank (equal to 1) and which is "well behaved" where M converges to the knot. 6(11) is a picture mnemonic for this notion. This essentially static structure on S^3 translates to a moving surface, tied to the knot like a sail, by the following bit of differential topology. Orient the circle on which f takes its values and parametrize it by degrees from 0 to 360 for typographical convenience. It follows from the implicit function theorem that the inverse image

$$F(t) = \{p \in M : f(p) = t\}$$

of each t is a properly embedded, two-sided, two-dimensional submanifold of M. The surfaces $F(t)$ are oriented by the gradient vector field grad(f), which is everywhere perpendicular to them. Choose one of these, $F = F(0)$, as reference surface and integrate along grad(f). This produces a flow on M that moves F through a succession of positions $F(t)$, with the last coinciding with the first. That is, for each point p in F, solve the initial value problem

$$\frac{dp}{dt} = \text{grad } f(p(t)) \, , \, p(0) = p \, .$$

Even though $F(360) = F(0)$, there is no reason to expect the flowlines to close up after one time around; in general $p(360) \neq p(0)$. You can find more about the knottedness of closed flowlines of grad(f) in Birman-Williams [1983]. The first-return map,

$$g : F \to F \, , \, g(p(0)) = p(360) \, ,$$

called the *monodromy* of the fibration, is a diffeomorphic self-map, or automorphism, of the surface. This is an example of a Poincaré map for the dynamical system. The manifold may be reassembled from the monodromy by identifying top and bottom of the cylinder $F \times [0,360]$ by the recipe $(p,0) = (g(p),360)$. A picture mnemonic for such a mapping torus $M(g)$ of the automorphism $g:F \to F$ might look like 6(12). The cross-sections of $M(g)$ looks like the conventional symbol for Seifert's surface as drawn in 6(21). It represents annuli joined together on a square. Such a ribbon neighborhood of two, 6(22), can be joined with a disc, 6(23), to form a whole torus. Thus F is a topological torus with a disc removed. The difference between 6(22) and all the other versions of Seifert's surface I have drawn is in the twisting of the ribbons. This is a property of the position of the surface in space, not of the surface as an abstract manifold.

Even if the monodromy is not the identity map on F, it is still possible that it is isotopic to it. That is, for a smooth map

$$h : F \times [0,1] \to F \, , \, h(p,0) = p \, , \, h(p,1) = g(p)$$

each $h|F \times \{t\}$ is a diffeomorphism. In that case the fibration is said to be trivial because $M(g)$ would be diffeomorphic to $F \times S^1$. As we shall see, this is not the case here. The isotopy class of g is an example of a so called *pseudo-Anosov* mapping in Thurston's [1976] classification of the automorphisms of surfaces. This is important, because by Jørgenson [1977], the mapping torus supports a hyperbolic geometry if and only if the monodromy is pseudo-Anosov.

THE OWL AND THE PUSSYCAT. *Figure 7.*

Here are some more pictures of F based on its definition as a disc with two oppositely twisted ribbons attached to its circumference. Unexpected equivalences of shapes result when a rubber sheet topologist twists ribbons. The

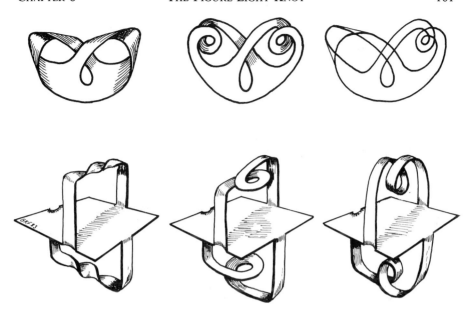

Figure 7   THE OWL AND THE PUSSYCAT

owl 7(12) proceeds from the pussycat 7(11) by a *Seifert move*, see Figure 7 of Seifert [1934], that changes the twisting of ribbons into curling or writhing. How their line patterns can be remembered from 6(21) is shown in 7(13). More useful, if less amusing, is the cubist position 7(21), a simpler form of which will concern us for the remainder of the chapter.

The importance of pictorial transmutations in the topology of surfaces, such as Seifert's move, warrants the following digression. Curl a strip of paper like the ribbons in 7(23) and pull it taut; it twists like 7(21). A ribbon perpendicular to the center line of the curl will writhe, like 7(22), but parallel to a vertical plane. Jim White explains the differential geometry of this phenomenon in Bauer-Crick-White [1980], where its natural manifestation in molecular biology is uncovered. Note how right and left handedness of twist, writhe and curl expresses itself. What remains invariant under an isotopy of a closed ribbon in space is the number of times one edge links the other. The number of obvious twists in the ribbon need not equal this integer; the difference is called its *writhing number*. Both twist and writhe are real numbers, not necessarily integers, associated with the geometry of a particular position in space, see White[1969]. However, for a picture of a closed ribbon in general position, Lou Kauffman [1985, Exercise 2.7] has a direct, visually convenient way of counting twists and curls. The first integer is the net number of times the ribbon twists, ignoring the global overpasses. The second integer is the number of times the ribbon passes over itself from left to right minus the number of times it does so from right to left, as seen from (either of) its directions. Their sum is the *linking number* of the ribbon edges.

ISOTOPIC SEIFERT SURFACES. *Figure 8.*

Of all the positions of K and an initial spanning surface F, 7(21) seems to be the most versatile, once the extraneous bends in K are removed by widening the bridges to form 8(11). To aid the eye, the knot is half striped, half cabled, and eight windows in F reveal otherwise invisible details. (Recall that a window is a transparent patch on the surface whose projection to the picture plane is a topological disc.) The surface is marked by two circles, which cross on the horizontal square floor and are directed towards the viewer by white and black arrowheads as they cross the bridges. This is position $F(0)$; $F(180)$ is shown below, 8(21). Here the horizontal floor extends away from the knot on the outside of the box-like frame. The generating circles of $F(180)$ cross at the "center" of this window through infinity. Two simple, closed, directed, once crossing curves on the Seifert surface (torus minus disc) are said to generate F because they determine the homeomorphisms of F up to isotopy. In other words, I only need to tell you where the generators go under a mapping in order to identify the isotopy class of the homeomorphism. To visualize the motion of $F(0)$ to $F(180)$, first imagine the floor of $F(0)$ moving upward, like an elevator, as t becomes positive. As the floor rises past a point on the knot, the surface turns 90° in the vicinity of this point. To see that the surface turns the same way all along the knot, orient the knot so that the striped part heads upward in back and downward in front. You should imagine the surface moving at other places as well, but so slowly that the picture does not change enough to require redrawing it. Soon, however, floor of $F(80)$ seems to get stuck below the white bridge.

CHAPTER 8 THE FIGURE EIGHT KNOT 163

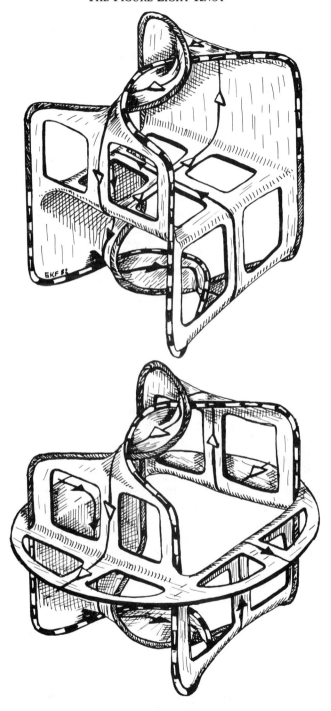

Figure 8 ISOTOPIC SEIFERT SURFACES

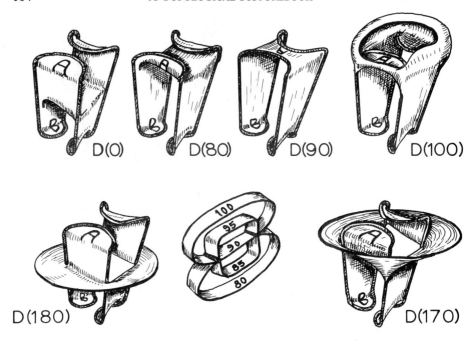

Trivial Fibration

Figure 9

Since the motion "through" a bridge will be the hardest part of the visual exercise, let's see what it would look like if the bridges were not there. This picture shows the relevant stages of a disc D spanning an unknotted curve C bent into a shape similar to the one at hand. At $F(90)$ the spanning disc nearly minimizes area. It bubbles out on top at $D(100)$. Note that during $80 < t < 100$ the surface swings almost 360° about the knot near A, while near B it remains almost stationary. The "almost" is important, because each $D(t)$ must be disjoint from every other. The bubble grows and encloses nearly all of the upper half space at $D(170)$. The balloon bursts at $D(180)$ as the disc passes through infinity. To imagine the rest of the tour, reflect the figures in the horizontal plane, turn them 90° and retrace your steps till $D(360)$ occupies the same position as $D(0)$. This motion describes the fibration over the circle by discs of the complement in S^3 of an unknotted, closed curve C.

Double Torus Knot. *Figure 10.*

Initially, I had many candidates for the position of the figure-8 knot and the Seifert surface spanning it. Here is an appealing one that fits on a double torus. There is no way it can fit on the surface of the standard torus. If it

Chapter 8 The Figure Eight Knot

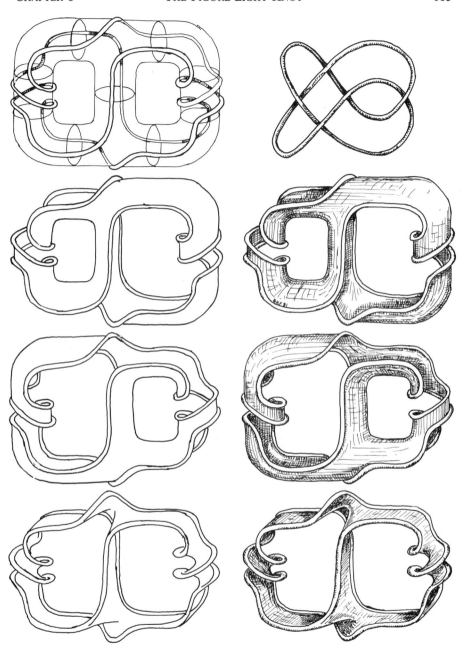

Figure 10   Double Torus Knot

did, there would be no story, for there is a direct way of visualizing fibrations of torus knot complements in S^3. For a complete and picturesque exploration of these matters, you should consult Rolfsen [1976]. In effect, the knot decomposes the double torus as a non-obvious connected sum of two "single" tori. The position of the knot on the double torus is shown in 10(11). This is also the line pattern for the two positions, 10(21) and 10(31). That you can move one of these to the other through the interior of the double torus is suggested by the midway position 10(41). Thus, it is easy to see half of the fibration, the one inside the double torus. To be sure, S^3 decomposes into the union of two such handle bodies and so one could wave one's hands at the other half. The blank space at the top is for the view of the surface as it passes through infinity. I leave this exercise to you.

TETRAHEDRAL EIGHT KNOT: ONE. *Figure 11.*

The figure-8 knot makes its entrance in Thurston's grand drama, fitted to the edges of a tetrahedron. The gluing diagram and many other properties are read from this position. 11(11) to 11(12) to 11(13) shows a transition from Seifert's position to Thurston's. Alternatively, put some corners into position 1(23). 11(21) and 11(22) show the two complementary positions of the spanning surface on the inside of the tetrahedron. The transition between these two is easy to see. Here, it is hard to see how to get the surface past the hooks. 11(31) shows what the imagination must interpolate between, at the vertical (right) hook of 11(21). 11(32) is an alternative view, with both pieces of 11(31) pushed together. This form is used in the next pair, 11(41) and 11(42). It is easy to see how to get from one side of the toroidal tube, 11(41), to the other, 11(42), by pushing. To see what happens in the T-joint, look to the right.

TETRAHEDRAL EIGHT KNOT: TWO. *Figure 12.*

A second tetrahedral position is given at 12(12). It corresponds to the drawing 2:13(21) which was used in Thurston's [1982] survey article, and by Kneale [1983]. At 12(32) is the line pattern on which all of the other views were based. To see the position of the knot on the double torus, 12(11), you must replace the two discs that were removed. Although these are not windows, their intended position is not hard to guess. The "average" position between 12(21) and 12(22) is given at 12(31). But the transition from 12(22) to 12(31) is not at all obvious.

Chapter 8 — The Figure Eight Knot

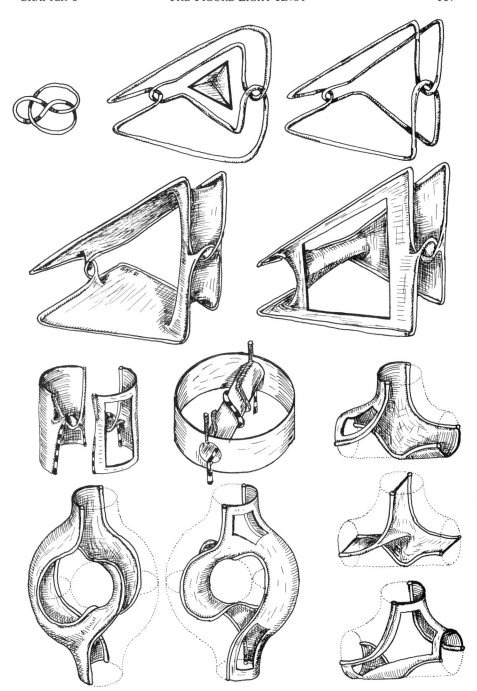

Figure 11 Tetrahedral Eight Knot One

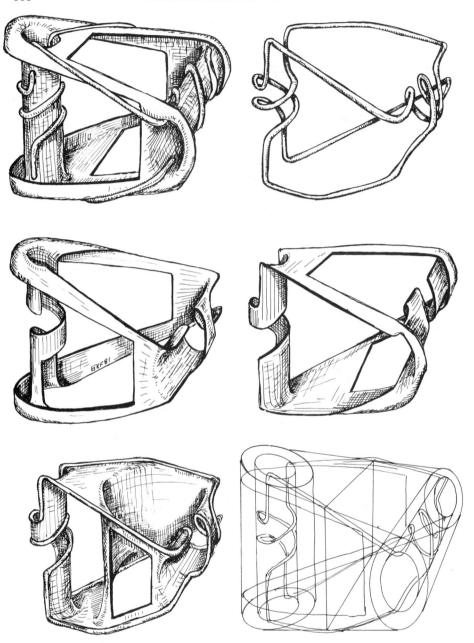

Figure 12   Tetrahedral Eight Knot Two

TOROIDAL SHEAR ISOTOPY. *Figure 13.*

By now, the solution to our visualization problem has narrowed down to two possibilities. A development of 11(41) and 11(42) is embodied in 13(21) and 13(23). The transition position, which helps the visualization, is at 13(22). But, my loyal and stalwart colleagues in the Geometric Potpourri found all these transitions too tedious to keep track of. Thus, somehow the transition between the two surface details of 11(31), repeated at 13(11) and 13(13), had to be made more explicit. What should 13(12) look like?

CYLINDRICAL SHEAR ISOTOPY. *Figure 14.*

This figure depicts five positions of an annulus A neighboring the white generator of Seifert's surface F as it squeezes through the upper bridge between $t = 80$ and $t = 100$. If these five $A(t)$ were superimposed into one drawing, it would correspond to 9(22). This detail of $F(t)$ was chosen to fit into 4(12) as well as into 8(11). At the bottom is $A(80)$ and $A(100)$ is at the top. The position midway between these, 14(22), is part of $F(90)$. Imagine $A(80)$ shrinking almost into its minimal shape, sliding freely about the two bent rods. To aid the imagination, I have also drawn a position for $t = 85$ on the left, 14(21), and $t = 95$ on the right, 14(23). The shrinking of $A(80)$ to $A(90)$ via $A(85)$ is not quite uniform. The bridge is somewhat stiffer than the rest, but catches up quickly between $A(85)$ and $A(90)$.

The annulus $A(80)$ is marked with an equatorial circle, white arrows, and a cross cut, black arrows. On $A(85)$ and $A(95)$ only the cross cut..., and on $A(90)$ only the equator has been drawn. You should trace out each of the missing curves to convince yourself that their position on $A(100)$ is as shown. The equator still crosses the bridge towards the front, just as on 8(21). The cross-cut, on the other hand, now turns right, following along the equator once around, before crossing it and continuing to the far edge of $A(100)$.

To express the global situation algebraically, let b and w denote the (isotopy classes of the) black and white generators of $F(0)$; B and W those of $F(180)$. Two simple closed curves on F are isotopic if you can slide one on top of the other without leaving the surface. Now draw a hemispherical dome over 8(21) that attaches to the circular rim of the floor. This approximates the position $F(100)$. As you can see from 14(11), $w = W$ but $b = B + W$. The symbol $B + W$ denotes the class of the closed curve obtained from a pair of generators of F by cutting them at their common point and rejoining them in the obvious way that does not produce conflicting directions on the new curve. This *switching operation* (Umschaltung) at crossings of directed curves and its higher dimensional analogue is one of the fundamental "cut and paste" procedures in topology. It was first used by Gauss in his study

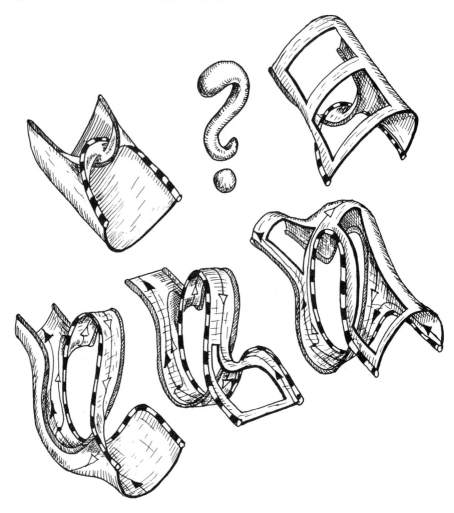

Figure 13   TOROIDAL SHEAR ISOTOPY

of knot diagrams and it is central to Seifert's method for detecting surfaces spanning knots. For the present, however, I only want to switch generators of F.

Thus, the generators (b,w) of $F(0)$ move to $(B+W, W)$ on $F(180)$. This suggests the name *shear isotopy* for the motion that squeezes a surface through a twisted bridge. A better reason for the name, whose full import must regrettably be omitted here, goes as follows. The surface F minus its border curve K may be regarded as a torus with a point removed. The punctured torus may, in turn, be parametrized by the plane R^2 minus the lattice Z^2 of points with integral coordinates. Points, whose coordinates differ

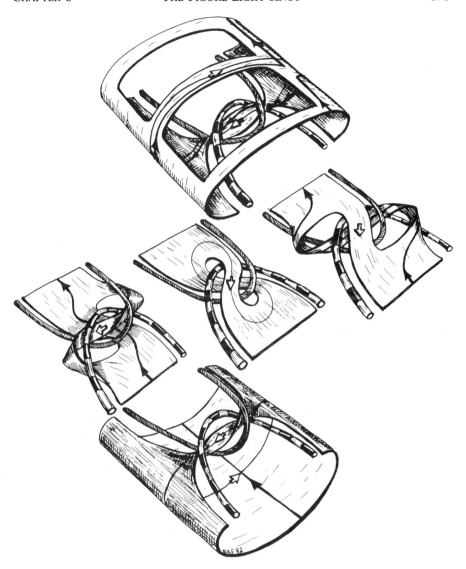

Figure 14 CYLINDRICAL SHEAR ISOTOPY

by an element of the translation group Z^2, name the same point on F-K. Up to isotopy, this self map restricted to $F - K$ may, this way, be represented by the linear shear transformation whose matrix is

$$\begin{vmatrix} 1 & 1 \\ 0 & 1 \end{vmatrix}.$$

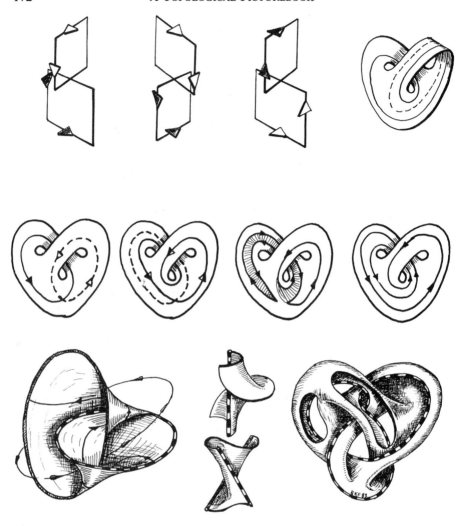

COMPUTING THE MONODROMY

Figure 15

To compute the fate of $(B + W, W)$ as $F(180)$ moves on to $F(360)$, it is convenient to reverse time and view the motion $F(360, 270, 180)$ through the spectacles of the turn-reflection. To see how this takes (b,w) to $(-w,b)$, turn 15(11) to 15(12) and reflect to obtain 15(13). (The curve $-w$ is w with direction reversed.) A ribbon twisting to the right reflects to one that twists to the left. Therefore, a picture of the shear isotopy $A(280, 270, 260)$ at the lower bridge would look just like $A(80, 90, 100)$, but with different markings. Therefore, this conjugation has the effect of taking

$$(b,w) \to (-w,b) \to (-W, B+W) \to (B, W-B).$$

So, since the forward motion from $F(180)$ to $F(360)$ takes $(B, W-B)$ to (b,w), it also takes $W = (W - B) + B$ to $w + b$, and therefore $B + W$ moves to $b + (w + b)$. Since the monodromy map induces an isomorphism of the first homology group $H_1(F)$ of the surface, the argument may be summarized algebraically by decomposing its matrix representation thus

$$\begin{vmatrix} 2 & 1 \\ 1 & 1 \end{vmatrix} = \begin{vmatrix} 1 & 1 \\ 0 & 1 \end{vmatrix} \cdot \begin{vmatrix} 0 & -1 \\ 1 & 0 \end{vmatrix} \cdot \begin{vmatrix} 1 & 1 \\ 0 & 1 \end{vmatrix} \cdot \begin{vmatrix} 0 & -1 \\ 1 & 1 \end{vmatrix}$$

These Gauss switches are legitimate because $b + w$ crosses each generator once, up to isotopy on F. This need not be the case on a one-sided surface, such as the Klein bottle, 15(14), with a disc removed. To see what happens on F, look at the "standard", untwisted version of Seifert's surface, 15(21), with generator b on the left, solid curve, and w on the right, dashed. The solid curve on 15(22) shows why $b + w = w + b$ and 15(31) shows how $(b + w) - b$ is isotopic to w. 15(23) shows the image of b under the monodromy, $g(b) = b + (b + w)$. You should trace into 15(24) a copy of $b + w = g(w)$ that crosses the given curve once, from left to right, showing that the monodromy is orientation preserving.

HOPF FIBRATION. *Figure 16.*

My picture story about the figure-8 knot is almost over, but there are a few loose ends to tie together. If you are familiar with the theory of fibered links in Stallings [1961, 1976], you may have noticed that grafting the shear isotopy and its mirror image into the trivial fibration is a local procedure, which specialist call "plumbing" two oppositely oriented "Hopf fibrations." See Harer [1982], for example.

I did not recognize this until some time after designing the shear isotopy. The first graft splits the unknot into a pair of circles which link once. If you close up the two bent rods in each detail of the shear isotopy into a pair of linked circles, L, then the successive positions of the annulus spanning the link, fibers $S^3 - L$ over S^1. If the second graft had also been left handed, the resulting Seifert surfaces, now embedded in S^3 with both straps 4(23) twisted alike, would fiber the complement of the trefoil knot. For a 3-crossing view of this knot, reverse "over" by "under" at the lower two crossings in 1(12) and twist the pendant loop, as in 1(21), directly.

Hopf's name is more commonly associated with the fibration of S^3 by circles over S^2 induced by the projection of the complex plane to the complex projective line. Any pair of circular fibers link. If you connect two of these linked circles by a one-parameter succession of them, 16(11), a once twisted annulus appears. 16(22) is a line pattern for the line drawing 16(32) which is shaded in at 16(42).

The monodromy of this fibration is just a Dehn twist, as you can detect from 15(31). One full clockwise turn of the striped rim of this annulus moves the marked cross-cut, $c = b\text{-}w$, off the twisted bridge. But the monodromy does the same thing, $g\,(b - w) = (b + w) - w = b$.

For more on this, see John Harer's [1982] account of how every fibered link in S^3, and even in an arbitrary 3-manifold, can be assembled out of such simple pieces. The plumbing of two Hopf links, as drawn in the square saddle shape of 7(21), illustrates a special case of a Murasugi sum. This generalized plumbing and its inverse operation, riginally treated algebraically, preserve several other geometric properties of link complements. Dave Gabai [1983] explores these geometrically in the first two of a series of papers on foliations and the topology of 3-manifolds. To him, perceiving how a spanning surface squeezes through unlikely places, as in the shear isotopy, is an exercise every low dimensional topologist should do.

Manipulating knot diagrams is really an exercise in two-and-a-half dimensions; one stays close to the picture plane for reasons of accuracy and convenience in coding the topological information. When truly spatial, rotatable forms are desired, for instance to fix in the imagination what the diagram represents, some simple, graphical tricks come in handy. How to attach a Seifert surface to the knot is the example at hand. A solid, tubular neighborhood, N, of K carries a canonical structure of simple closed curves on its toroidal surface T, called a *framing*. The meridians, which go around T the short way, span discs inside N. Of all possible curves that go once around T the long way, only the so called longitudes span a 2-sided surface in the knot exterior, $X = S^3 - N$. The Meyer-Vietoris isomorphism,

$$H_1(T) = H_1(X) + H_1(N) ,$$

characterizes these curves homologically: $(1,0)$ is the class of meridians, $(0,1)$ is that of the longitudes. Thus Seifert's surface attaches to T along a longitude, and a longitude does not link the knot.

Now from any view, the two contours of T visibly follow once around K the long way, but they need not be longitudes. Spanning surfaces are easier to draw when they are. I used elementary cusp forms, 15(32)), to slide 4(12) around the knot until F attaches to T entirely along one contour, 15(33). This led to position 14(22), midway between 14(31) and 14(11) in the shear isotopy.

Chapter 8 — The Figure Eight Knot

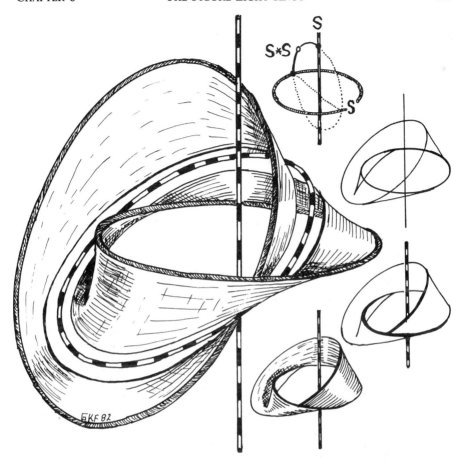

Figure 16 Hopf Fibration

POSTSCRIPT

In the nineteenth century, photographers first applied their new medium to inexpensive substitutes for standard art: they produced landscapes and portraits. Later came motion pictures — something only the camera could do. The goal of much computer graphics today is still to produce electronic images of photographic quality or videotapes and disk-stored sequences of precomputed movies. Of greater use for descriptive topology would be an *inexpensive* graphics tool for producing *recognizable* images by *oneself*, in huge *quantities* and *fast* enough to keep from wasting time. As yet, we have to strike a compromise on each of these five requirements.

Over the decade during which the pictures in this book were designed, drawn, and redrawn, interactive computer graphics became increasingly available also to the journeyman topologist. I am often asked whether my pictures were done on a computer, and even if they weren't, could they have been? Until recently I confidently answered both questions with a "no." This summer a project applying some of the techniques developed herein to computer graphics bore the first fruits. At the same time, it turned out that the majority of my photographs of long-erased blackboard pictures were not good enough for color printing. Film processing specialist Rich Becker was able to rescue seven of them. Two color plates now show some representative frames from mouse-controlled computer animations. Such user-controlled "movies" are uniquely possible on a computer.

Donna Cox, a computer artist in the School of Art and Design, has a project on the convergence of art and science that uses the graphics facilities of the National Center for Supercomputing Applications (NCSA), directed by Larry Smarr and funded by the National Science Foundation. She is also checking out (beta-testing) a solid geometry modeler, the Cubicomp PICTUREMAKER in the Electronic Imaging Lab. A solid geometry modeling system is designed to allow the artist first to construct various box-like shapes with flat faces having arbitrary polygonal borders. Such *facets* are oriented and each side can separately be defined to be invisible, semi-transparent or opaque. In the latter case, the application of color, called *rendering*, can be so blended as to give the illusion of continuous curvature. Under mouse-control, the solids can be resized, repainted, rotated, translated, and permanently attached to each other to make more complex objects, on which this process can be repeated.

In a demonstration of how such a program can also be used to construct the cross cap by the process described on page 87, Donna first mouse-drafted circles and ellipses in a plane. It is impractical to give the Cubicomp mathematical formulas even for such simple curves. On the other hand, the program does automatically interpolate the facets between two elliptic discs that have been moved near each other, forming a cylindrical belt between them. The polyhedra are then positioned and permanently attached to each other. The first three pictures give an equatorial view III(11), a polar view III(13) and one roughly like 5:4(23) in between. The color coded facets were left transparent to reveal the structure of the model. With opaque facets of one color and with some specular reflections painted in, but not yet with continuous curvature, the cross cap could look like the mystery object III(14) on the right. The limitations of solid geometry modelers are well illustrated here. Can you figure out from just one picture how III(14) was made?

The remaining computer generated frames on color plates were all drawn on a Silicon Graphics IRIS 2400 and photographed by a Dunn Instruments 635 Film Recorder at the NCSA. Half-again as expensive as the Cubicomp and worth nearly 100 Apples, the IRIS is UNIX controlled and therefore very easy to program in both C and FORTRAN. I originally worked out the geometrical part of the C program in Isys FORTH. This implementation of Charles Moore's language on the Apple was created by Bob Illyes [1984]. It optimizes the graphics capbility of the Apple's modest 6502 chip.

With Illyes' help in mastering this wonderfully mathematical computer language, I was able to explore and improve many geometrical programs intended for bigger machines. Modifications of Apery's Romboy homotopy (see p. 96) proved particularly apt for Donna Cox's project. On the IRIS, images composed of individual colored pixels, such as III(24) and III(34) can be rotated with a slide of the mouse. Our programmer and hardware specialist, Ray Idaszak, invented a simple, quick and remarkably satisfactory coloring algorithm which renders such a *pixel cloud* in 40 seconds at the click of a button.

First, Idaszak assigns 1-color Lambert shading (see pp. 60-64) to the triangular facets associated with adjacent, not necessarily co-planar quadruples of points on a surface. Thus views III(21), III(22), and III(23) of a yellow cross cap correspond to the positions immediately above. For the complementary colors, which added to yellow make white, the intensity is proportional to a very high power of Lambert's cosine. Thus those facets facing the light source are rendered very brightly, giving the surface highlights more or less in the right place. The facets themselves are made smaller than a pixel, and therefore the surfaces look reasonably smooth without blending.

These ingenious improvements on otherwise standard techniques prevent fatal rendering errors common programs are prone to make here. These errors are due to the fact that the surface penetrates itself along double curves and reverses orientation near pinch points.

For a unified description of the geometry expressed by these images, reconsider Apery's cylindrical parametrization of the Roman surface (p. 96). For each equatorial angle θ the conjugate axes of the ellipse are the columns of a 2×3 matrix $L(\theta)=\langle J(\theta), K(\theta)\rangle$ considered as a linear mapping of a plane into space. The pair of sliding coefficients are the Cartesian coordinates of a circle $C(t)=\langle A(t), B(t)\rangle$ through the origin, whose polar equation is $\rho=\cos(\tau)$, where $\tau=\arctan(t)$. It is visually preferable to let τ measure the declination from the vertical. Thus $A(t)$ becomes the "altitudinal" coordinate and $B(t)$ the "basal" coordinate of a circle resting on a horizontal line.

Apery's Romboy homotopy consists in shearing this circle into a long, skinny ellipse:

$$\rho=\cos(\tau)/(1-\beta\sin(2\tau)), \text{ where } \beta=(b/\sqrt{2})\sin(3\theta).$$

The pinch points cancel at $b=1/\sqrt{3}$ producing Boy's immersion of the projective plane in 3-space. The circle $\rho=\cos(\tau)$ is described twice as τ turns once through 360°. Suppose we *uncurl* it to the unit circle by a linear homotopy given in polar coordinates by $\rho=(1-\ell)\cos(\tau)+\ell$. This forms the *limaçons of Pascal*; see Lawrence [1972, p. 113]. In the first half of the deformation, the limaçon is a double loop, with the node at the origin. At $\ell=\frac{1}{3}$, it is *Pascal's trisectrix*. The smaller loop shrinks to the cusp of the *cardioid* at $\ell=\frac{1}{2}$ as in 5:6(21); see page 93. The dimple finally pops out to form the unit circle at $\ell=1$.

The uncurled Roman surface, shown at III(44) with a sinusoidal color map, is swept out by the oval image of the unit circle under the affine maps $L(\theta)$. Let me call it the *Etruscan surface* because it is computationally simpler than the Roman surface. Topologically, the Etruscan surface is a singular Klein bottle obtained by performing the connected sum of two Roman surfaces at their south poles. Interesting new surfaces, generated by limaçons, appear along the way. The pixel clouds show the apparent contour of the surface generated by limaçons near the trisectrix stage. The same surface as seen from the south pole is given by III(43). Idaszak's algorithm draws the succession of bands between adjacent ovals and can thus be interrupted to reveal additional detail, as in III(41) and III(42).

Many other surfaces have an *ovalesque* parametrization. For example, the cross cap can be generated by ellipses with perpendicular conjugate axes:

$$J(\theta)=\langle \cos(2\theta), \sin(2\theta), 0 \rangle$$
$$K(\theta)=\langle 0, 0, \cos(\theta)\rangle.$$

Images III(31), III(32) and III(33) depict three views, corresponding to those above, of the cross cap uncurled to the trisectrix stage.

One remarkable and totally unexpected view of the Etruscan surface,

IV(23), prompted the nickname "Venus de Iris". Since the product of the two homotopies, Romboy and Limaçon, could be installed in the program at no extra expense, Ray was able to create the immersed Klein bottle, IV(14), nicknamed "Ida". The polar equation of the 3-parameter family of quartic plane curves so produced is

$$\rho = \frac{(1-\ell)\cos(\tau)+\ell}{1-b/\sqrt{2}\,\sin(3\theta)\sin(2\tau)}$$

where θ is coupled with the 1-parameter family of affine maps

$$L(\theta) = \begin{vmatrix} r_1 \cos(2\theta) & \cos(\theta) \\ r_1 \sin(2\theta) & -\sin(\theta) \\ r_2 & 0 \end{vmatrix}$$

which sweep out the surfaces, call them $F(\ell ITb)$, shown on the fourth color plate. The waist to height ratio, r_1/r_2, of the Venus, $F(1,0)$ at IV(44), is close to ½. A side view of the Roman surface $F(0,0)$ is in the corner IV(41). The Boy surface at IV(11) is $F(0,1)$. The transitional forms IV(21) and IV(31) flank the pinch point cancellation of the Romboy homotopy. Similarly the transition shapes IV(42) and IV(43) flank the surface $F(½,0)$ generated by cardioid limaçons. The corresponding shapes obtained by uncurling Boy's surface are in the top row, while 12 pinch points of the Venus cancel somewhere between IV(34) and IV(24) on the way to Ida, $F(1,1)$ at IV(14). Thus Apery's implementation of Morin's pinch point cancellation procedure appears to be quite general. The 3 green surfaces at IV(33), IV(32) and IV(22) are rotations of IV(44), IV(34) and IV(24) respectively.

Mathematical typography has also much improved since I began this manuscript with *Applewriter* by Paul Lutus [1981]. A rare feature of this little word processor, its programmability, became indispensible when my electronic manuscript was modified for telephone transmission to the typesetting computer of the University of Illinois Printshop. Computer drafting is another promising tool for descriptive topology. For the sake of comparison with hand drafting, I made the upper half of figure 2:1, page 15, on a MacIntosh. The cubic curve is a spline from Mark Cutter's *MacDraw* [1985], the lettering is from Atkinson's *MacPaint* [1985], but rotated by a co-resident program, *ClickArt* by Bill Parkhurst [1985]. This graphic was printed on a laser printer, while the table on page 129, also a MacPaint picture, was made on a more modest dot-matrix printer.

Clearly yesterday's "heroic" efforts in bringing computers to descriptive topology (see p. 29) will become routine sooner than I thought, thanks to the increasingly more realistic attitude of financial granting agencies towards funding non-commercial computer graphics on the campus.

BIBLIOGRAPHY

The braces contain the locations in this book where the entry has been cited.

Apery, François: La Surface de Boy. Dissertation, Louis Pasteur University, Strasbourg, 1984. {77,96}

Abikoff, William. *The Real Analytic Theory of Teichmüller Space.* Lect. Notes in Math. 820. Springer-Verlag, Berlin-Heidelberg-New York, 1980 {Preface}

Abikoff, William. The Uniformization Theorem. *Amer. Math. Monthly* **88**(1981), 574-592. {Preface}

Abraham, Ralph and Christopher Shaw: *Dynamics–the Geometry of Behavior.* Aerial Press, Inc., Santa Cruz, CA, 1982.

Atkinson, Bill: *MacPaint.* Apple Computer, Inc., Cupertino, CA, 1985. {179}

Bauer, W., F. Crick and J. White: Supercoiled DNA. *Scientific American* **243**(July 1980), 118-133. {161}

Banchoff, Thomas, Terence Gaffney and Clint McCrory: *Cusps of Gauss Mappings.* Res. Notes in Math. 55, Pitman Books Ltd., London, 1982. {89,106}

Banchoff, Thomas and Charles Strauss: *Complex Function Graphs, Dupin Cyclides, Gauss Map, and Veronese Surface.* Computer Geometry Films. Brown University, Providence 1977. {29,89}

Berstein, Israel and Allan Edmonds: On the construction of branched coverings of low-dimensional manifolds. *Trans. Amer. Math. Soc.* **247**(1979), 87-124. {Preface}

Birman, Joan: *Braids, Links and Mapping Class Groups.* Ann. Math. Studies 82, Princeton Univ. Press, Princeton, 1974. {75,129,138,143}

Birman, Joan and Robert F. Williams: Knotted periodic orbits in dynamical systems-I: Lorenz' Equations. *Topology* **22**(1983), 47-82; -II: Knot holders for fibered knots. In *Symposia Mat.* **XI**, Academic Press, 1983. {160}

Boy, Werner: Über die *Curvatura integra* und die Topologie geschlossener Flächen. Dissertation, Göttingen, 1901. *Math. Ann.* **57**(1903),151-184. {77}

Boyer, Carl: *The History of the Calculus and its Conceptual Development.* 1949. Reprinted by Dover Publications Inc., New York, 1959. {14}

Bröcker, Theodor and Les Lander: *Differentiable Germs and Catastrophes*. London Math. Soc. Lect. Notes 17. Cambridge Univ. Press, Cambridge, 1975.   {Preface}

Burmester, Ludwig: *Theorie und Darstellung der Beleuchtung Gesetzmässig Gestalteter Flächen*. Teubner, Leipzig, 1875.

Callahan, James: Singularities of plane maps. *Amer. Math. Monthly* **81**(1974), 211-240, and **84**(1977), 765-803.   {98}

Cohen, Marshall: Whitehead torsion, group extensions, and Zeeman's conjecture in high dimensions. *Topology* **16**(1977), 79-88.   {22}

Cowan, Thaddeus: The theory of braids and the analysis of impossible figures. *J. Math. Psychol.* **11**(1974), 120-212.   {67}

Cremona, Luigi: Sulle transformazione geometriche delle figure piane. *Gior. d. Mat.* **1**(1863), 305-311, and **3**(1865), 269-280.

Crowell, R. H. and R. H. Fox: *Introductin to Knot Theory*. Ginn & Co., Boston, 1963.   {28}

Cutter, Mark: *MacDraw*. Apple Computer, Inc., Cupertino, CA, 1984.   {179}

Dehn, Max: Die Gruppe der Abbildungsklasse. *Acta Math.* **69**(1938), 135-206.
   {142}

Doblin, Jay: *Perspective: a New System for Designers*. Whitney Library of Design, New York, 1956.   {46,56}

Dyck, Walter von: Analysis situs I. *Math. Ann.* **32**(1888), 457-512.   {101,116}

Fadell, E. and J. VanBuskirk: The braid groups of E^2 and S^2 *Duke Math. J.* **29**(1962), 243-258.   {136}

Francis, George K. and Stephanie Troyer: Excellent maps with given folds and cusps in the plane. *Houston J. Math.* **3**(1977), 165-194, Continuation. *Ibid.* **8**(1982), 53-59.   {110,117,147}

Francis, George K. and Bernard Morin: Arnold Shapiro's eversion of the sphere. *Math. Intelligencer* **2**(1979), 200-203.   {104,107}

Francis, George K.: Drawing surfaces and their deformations: the tobacco pouch eversions of the sphere. *Math. Modelling* **1**(1980), 273-281.   {109}

Francis, George K.: Drawing Seifert surfaces that fiber the figure-8 knot complement. *Amer. Math. Monthly* **90**(1983), 589-599.   {Preface}

Freedman, Michael: The topology of four-dimensional manifolds. *J. Diff. Geom.* **17**(1982), 357-452.   {P82}

French, Thomas: *A Manual of Engineering Drawing for Students and Draftsmen*. 1911, Seventh Edition, McGraw-Hill, New York, 1947.   {43}

Gabai, David: Foliations and the topology of 3-manifolds. *J. Diff. Geom.* **18**(1983), 445-503.   {174}

Gabai, David: The Murasugi sum is a natural geometric operation. In *Proc. Conf. on Low Dim. Top., Contemp. Math.* **20**(1983), 131-143.   {174}

Gillman, D. and Dale Rolfsen: The Zeeman conjecture for standard spines is equivalent to the Poincaré conjecture. *Topology* **22**(1983), 315-323.   {22}

Golubistky, Martin and Victor Guillemin: *Stable Mappings and Their Singularities*.

Grad. Texts in Math. 14, Springer-Verlag, New York-Heidelberg-Berlin, 1973.
{1,8}

Gombrich, E. H. and R. L. Gregory, editors: *Illusion in Nature and Art*. Charles Scribner's Sons, New York, 1973. {62}

Gordon, McA. C.: Some aspects of classical knot theory. In *Knot Theory, Proceedings Plans-sur-Bex, 1977*. Lect. Notes Math. 685, Springer-Verlag, Berlin-Heidelberg-New York, 1978.

Grimm, Brüder:*Sechs Märchen-Bilderbücher mit Bildern von Brünhild Schlötter*. Verlag Jos. Scholz, Mainz. Gebunden von Jos. Joh. Schweizer, Leitmeritz, 1942.
{Preface}

Harer, John: How to construct all fibered knots and links. *Topology* **21**(1982),263-280. {173}

Hatcher, Allan and William Thurston: A presentation for the mapping class group of a closed orientable surface. *Topology* **19**(1980), 221-237. {144,148,147}

Hilbert, David and S. Cohn-Vossen: *Anschauliche Geometrie*. Springer-Verlag, Berlin, 1932. {86,89}

Hopf, Heinz: *Differential Geometry in the Large*. Lect. Notes. Math. 1000, Springer-Verlag, Berlin-Heidelberg-New York, 1983. {98}

Illyes, Robert: *Isys FORTH*. Illyes Systems, Champaign, IL, 1984. {177}

Jørgenson, Troels: Compact 3-manifolds of constant negative curvature fibering over the circle. *Ann. Math.* **106**(1977), 61-72. {150,160}

Kauffman, Louis: *On Knots*. U. Illinois at Chicago preprint, 1985. {161}

Kim, Scott: An impossible four-dimensional illusion. In *Hypergraphics*, edited by D. W. Brisson, 1978, Chapter 11, 186-239. {62,68}

Kneale, Dennis: Shaping Ideas: A Topologist Wows World of Math by Seeing the Unseen. *Wall Street Journal*, March 18, 1983. {36,166}

Kronecker, Leopold: Über Systeme von Funktionen mehrer Variablen. *Monatsber. Berl. Ak. Wiss.*(1869), 688-689. {101,103}

Krylov, N., P. Lobandievsky and S. Men: *Descriptive Geometry*. MIR Publishers, Moskow, 1968. {61}

Lawrence, J.Dennis: *A Catalog of Special Plane Curves*. Dover Publications, Inc., New York, 1972. {178}

Lickorish, W. B. Raymond: A finite set of generators for the homeotopy group of a 2-manifold. *Proc. Cambridge Philos. Soc.* **60**(1964), 769-778. {142}

Listing, Johann: Vorstudien zur Topologie. *Göttinger Studien* (1847),811-875.
{150}

Lutus, Paul and Liane Finstad: *Apple Writer II*. Apple Computer, Inc., Cupertino, CA, 1981 {179}

Lyndon, Roger and Paul Schupp: *Combinatorial Group Theory*. Ergebnisse der Mathematik 89, Springer-Verlag, Berlin-Heidelberg-New York, 1976. {72}

Mach, Ernst: *Beiträge zur Analyse der Empfindungen* Jena, 1886. *The Analysis of Sensation* Dover reprint, 1959. {61}

Magnus, Wilhelm: Über Automorphismen von Fundamentalgruppen berandeter Flächen. *Math. Ann.* **109**(1934), 617-646. In *Collected Papers* pages 67-96. {134}

Magnus, Wilhelm: Braids and Riemann surfaces. *Comm. Pure Appl. Math.* **25**(1972), 151-161. In *Collected Papers* pages 573-583. {172}

Magnus, Wilhelm: *Noneuclidean Tesselations and Their Groups.* Academic Press, New York, 1974. {150}

Magnus, Wilhelm: *Collected Papers.* Springer-Verlag, New York-Berlin-Heidelberg-Tokyo, 1984.

Marin, André: Seminaire sur les diffeomorphisms des surfaces d'apres Thurston. Exposé 11-12, *Inst. Haute Études Sci.*, 1977. {144}

Massey, William: Proof of a conjecture of Whitney. *Pacific J. Math.* **31**(1969), 143-156. {95}

Max, Nelson: *Turning a Sphere Inside Out.* International Film Bureau, Chicago, 1977. {109}

Max, Nelson and Thomas Banchoff: Every sphere eversion has a quadruple point. In *Contributions to Analysis and Geometry (Philip Hartman Festschrift).* Johns Hopkins University Press, Baltimore, 1981, pages 191-209. {123}

Milnor, John: Hyperbolic geometry: the first 150 years. *Bull. Amer. Math. Soc.* **6**(1982), 9-24. {149}

Möbius, August : Theorie der elementaren Verwandschaft. *Werke,* vol.2. Martin Sandig Verlag, Wiesbaden 1967. {77,82}

Morin, Bernard: Equations du retournement de la sphere. *C.R.A.S.P.* **287**(1978).
{98,116}

Morin, Bernard and Jean-Pierre Petit: Le retournement de la sphere. *Pour la Science* **15**(1979), 34-41. {108}

Nagase, Teruo: On elementary deformations of nice immersions of a surface into a 3-manifold, Tokai University preprint, 1984. {104}

Newman, M.H.A.: On a string problem of Dirac. *J. London Math. Soc.* **17**(1942), 173-177. {134}

Nicolaïdes, Kimon: *The Natural Way to Draw.* Houghton Mifflin, Boston, 1975.
{64}

Nielsen, J.: Untersuchungen zur Topologie der geschlossenen zweiseitigen Flächen I,II,III. *Acta Math.* **50**(1927), 189-358, **53**(1928), 1-76, **58**(1931), 87-167. {126}

Nitsche, Johannes: *Vorlesungen über Minimalflächen.* Springer-Verlag, Berlin-Heidelberg-New York, 1975. {119}

Penrose, L. S. and Roger Penrose: Impossible objects: a special type of illusion. *Brit. J. Psychol.* **31**(1958), 31-33. {65}

Petit, Jean-Pierre and Jérôme Souriau: La Surface de Boy. *La Recherche*, April 1982. {96}

Phillips, Anthony: Turning a surface inside out. *Scientific American* **214**(May 1966), 112-120. {108}

Poincaré, Henri: Analysis Situs. *J. d'Ecole Polytechnique Normale* **1**(1895), 1-121.
{125}

Poston, Tim and Ian Stewart: *Catastrophe Theory and its Applications.* Pitman Books Inc., London, 1978. {11}

Riley, Robert: A quadratic parabolic group. *Math. Proc. Camb. Philos. Soc.* **77**(1975), 281-289. {150}

Rohn, Karl and Erwin Papperitz: *Lehrbuch der Darstellenden Geometrie.* Verlag von Veit, Leipzig, 1906. {61}

Rolfsen, Dale: *Knots and Links.* Publish or Perish, Inc., Berkeley, 1976. {166}

Rotman, Joseph: *An Introduction to the Theory of Groups.* Allyn & Bacon Inc., Boston, 1984. {22}

Schoenflies, Artur: *Einführung in die Hauptgesetze der Zeichnerischen Darstellungsmethoden.* Teubner, Leipzig, 1908. {Preface}

Scholz, Erhard: *Geschichte des Mannigfaltigkeitsbegriffs von Riemann bis Poincaré.* Birkhäuser, Boston-Basel-Stuttgart, 1980. {89}

Seifert, Herbert: Topologie dreidimensionaler gefaserter Räume. *Acta Math.* **60**(1933),147-238. {150}

Seifert, Herbert: Über das Geschlecht von Knoten. *Math. Annalen* **110**(1934),571-592. {151,155,161}

Smale, Stephen: A classification of immersions of the two-sphere. *Trans. Amer. Math. Soc.* **90**(1959), 879-882. {61}

Smale, Stephen: Generalized Poincaré Conjecture in dimension greater than 4. *Ann. Math.* **74**(1961), 391-406. {125}

Stallings, John: On fibering certain 3-manifolds. In *Topology of 3-Manifolds and related topics*, edited by M. K. Fort, Jr., Prentice-Hall Inc.,Englewood Cliffs, 1962. {173}

Stallings, John: Construction of fibered knots and links. In *Symposium on Algebraic and Geometric Topology, Stanford, 1976.* A.M.S. Proc. Symp. Pure Math. **32**(1978),55-60. {173}

Steiner, Jakob: Zwei specielle Flächen vierter Ordnung. Posthumous report and editorial note number 32 on pages 721-724, 741-742 of *Gesammelte Werke. Zweiter Band.* K. Weierstrass, editor. Druck und Verlag von G. Reimer, Berlin, 1882.

Stillwell, John: *Classical Topology and Combinatorial Group Theory.* Grad. Texts in Math. 72, Springer-Verlag, New York-Heidelberg-Berlin, 1980. {129}

Thom, René: *Stabilité structurelle et morphogénèse.* W. A. Benjamin, Inc., Reading, Mass., 1972. {Preface}

Thurston, William: On the geometry and dynamics of diffeomorphisms of surfaces. Princeton University preprint, 1976. {160}

Thurston, William: *The Geometry and Topology of Three-Manifolds.* Princeton University preprint, 1977 and 1982. {149,151,157}

Thurston, William: Three dimensional manifolds, Kleinian groups, and hyperbolic geometry. *Bull. Amer. Math. Soc.* **6**(1982),357-381. {36,149}

Thurston, William: Hyperbolic structures on 3-manifolds, I: Deformation of acylindrical manifolds. *Annals Math.* **124**. (1986),203-246. {36,40}

Waddington, C. H., editor: *Towards a Theoretical Biology.* Edinburgh Univ. Press, 1972.

Wajnryb, Bronislaw: A simple presenation for the mapping class group of an orientable surface, Technion preprint, Haifa 1982. {144}

Wiener, Christian: *Lehrbuch der Darstellenden Geometrie*. Teubner, Leipzig, 1884.
{63}

White, James: Self-linking and the Gauss integral in higher dimensions. *Amer. J. Math.* (1969), 693-728. {68,161}

Whitney, Hassler: On regular closed curves in the plane. *Compos. Math.* **4**(1937), 276-284. {105}

Whitney, Hassler: On the topology of differentiable manifolds. In *Lectures in Topology*. Univ. Michigan Press, Ann Arbor, 1941. {92}

Whitney, Hassler: On singularities of mappings of Euclidean spaces. *Ann. Math.* **62**(1955), 374-410. {Preface}

Wolf, Joseph: *Spaces of Constant Curvature*. Publish or Perish Inc., Berkeley, 1977. {74}

Zeeman, E. C.: On the dunce hat. *Topology* **2**(1964), 341-358. {22}

Zeeman, E. C.: A catastrophe machine. In *Towards a Theoretical Biology*, edited by C. H. Waddington, 1972. In *Catastrophe Theory, Selected Papers 1972-1977*, pages 409-415. {120}

Zeeman, E. C.: *Catastrophe Theory, Selected Papers 1972-1977*. Addison-Wesley, Reading, Mass., 1977.

INDEX

Italic numbers refer to pages with pictures of the entry. Roman numbers refer to color plates. See also page references to authors in the Bibliography.

A

Affine projection 20
Apple computer 98
Abus de dessin 80,82,122
Amplitudo 105
Anastrophe 16
Art gum eraser 25,86
Artmann, Benno
  Darmstadt seminar 28
Apery, François
  Morin student 78
  Romboy homotopy 96,177
  Boy surface equation 98
Asimov, Dan
  Sudanese Möbius Band 24
Astroid, *see* curve
Automorphism 72

B

Backlight 63
Banchoff, Thomas
  computer geometry films 29
Baseball move 107
BASIC 31,63,98
Bifurcation 11
Birman, Joan
  favorite involution 136-138,146
Blague, see tobacco pouch
Blackboards 26,176
Blackmore, Colin
  Mach line illusion 62

Border
  curve 16
  profiling 16
  sum 102
Bott, Raoul
  asks to see eversion 105,117
Boy, Werner
  projective plane 89
  regular homotopy 104
Boyer, Carl
  calculus history 14
Braid *130,131-132*
  diagram 131
  product 131
  spherical 131
  strands 131
Bridge, twisted *156,158,163,167*,169-170,*171*
Buguer, Pierre
  illumination theory 61

C

C 177
Cancelling pinch points
  31-34,*32,34*,95,*103*,104,178
Catastrophe
  elliptic umbilic 13
  hyperbolic umbilic 98
  machine 120,*121*
  theory 14
Central extension 136
Chapeau 104,*108*,112

INDEX

Characteristic class 92
Chimera 82
Circle
  cut 148
  helper 144
  indicator 138
  longitude 128
  meridian 128,174
  obvious 137-139,146
  topological 128
  twelve-point 57
  twisting 138,146
Coil *133*
  braid 134
  matrix 146
  relator 129
  self-map 136
Cohen, Marshall
  dunce hat construction 22
Coin, box shape
  105,106-107,121-122,153
Collapsing 21-22
Collar 139
Computer
  Apple 96,98,177
  animation 109,176
  Cubicomp 177
  graphics 29,176
  IRIS 177
  micro 31
  PLATO 35
  word processor 179
Conjugate
  axes 96,178
  harmonic 48
Connected sum 102,178
Contour
  apparent fold 9
  astroid shaped 120
  estimate location 2,4,7,80
Corkscrew 131
Covering
  branched 138
  double 106
  universal 74
Cox, Donna
  computer artist 176-177
Cremona, Luigi
  analytic geometry 77
Cross-ratio function 48
Cube 50
Curvature
  drawing 26,177
  Gaussian 87

  integral 101,103
  mean 87
  positive 86
  negative 40,86
  sectional 87
Curve
  astroid *109*,116,117,120
  cardioid 178
  contour 8-9,16,80,*109,111*
  border 16
  deltoid 13,91,*109*,116
  double 8,16,33,112
  generating 2
  latitude 14
  limaçon 178
  longitude 107
  meridian 106
  parabolic 86
  perspective parabola 78
  perspective circle 56
  regular homotopy 91
  shading guide 18
  tangent envelope 116
  trisectrix 178
Cusp
  Cayley 11,24,*110*
  point 16
  rolling *90*,91,96,*97*,98
  saddle 11,*30*
Cut
  cross 122,139
  system 148

D

Decorations 15,*34*
Deltoid,*see* curve
Deformation
  completely continuous 103
  Dehn twist 122,128,138,144
  elevate motions 114
  limaçon 179
  rolling cusps 96-98
  Romboy 95-96,178-179
  uncurling 178-179
  see also
  cancelling pinch points
  homotopy
Descriptive geometry 1,45
Descriptive topology
  computer assisted 29,176
  elements listed 12
  like industrial design 45
  meets Bott's challenge 117

Descriptive Topology *cont.*
  mission defined 1
  Morin's method 78
Determinant
  Hessian 88
  Jacobian 86,87,116
Diagram
  braid 131
  flat drawing 43,78
  gluing 150-151
  schematic 149
Dihedral angle 151
Direction
  temporal 92
  chromatic 92
Dirac phenomenon 134
Disc
  diameter 151
Double
  a cube 54,*55*
  line *7*,8
  point 8
  torus 126,138,164,166
Doppelring 126
Doblin, Jay
  perspective method 56
Drawing
  chalk 26,I,II
  computer 29,*30,32,III,IV*
  computer assisted 33-34,37-39,175
  flat, *see* diagram
  line *15*,16,*17,33,40-42*
  perspective 44
  proofreading errors 108
  shaded 15
  tools 25,176
Drawing style
  Escher 64
  Phillips 108-109
  Petit 109
  Poston 11,*35*,36
  Scientific American 108
Drehspiegelung 151
Dürer, Albrecht
  teaching perspective 44
Duns
  egg 22,*23*,27
  Scotus 20
Dunce hat 20,*21*

E

Earphones, *see* surface

Edges *21*
  contour 18
  face 18,20
Electronic
  images 176
  manuscript 176
Elevation
  into third dimension 114
  orthographic view 43
Escher, Maurits
  impossible figures 20,65
  perspective experiments 64
Euclidean
  algorithm 99
  space forms 68
Eversion
  Max 29,114
  Morin 29,112,120
  Petit 108
  Phillips 108
  Shapiro 105-106
  spherical 29,98,99
  tobacco pouch 109,*115*,120
Euler characteristic 69,102
Eversion of spheres 98

F

Facet 176-177
Fibration
  fiber bundle 26
  fibered links 172
  Hopf 150,173
  over a circle 159
  singular 150
  Stallings 159,164
  trivial 160,164
Fine arts 45
Focal
  center 46
  disc 46
  distance 44,46,54
  set 106
Focus 48
Fold
  apparent 9
  shading 64
  singularity 9
FORTRAN 177
Framing
  knots 174
  line pattern 14
  triangle 49,*51*
French, Thomas
  curve 26

G

Gastrula shape 114
Gauss, Carl Friedrich
  amplitudo 104
  curvatura integra 103
  map 87,*103*
  switch 173,169
General position 1,16,27
Generators
  handle 139,143,148
  mapping class group 142
  Seifert surface 162
  surface 3
Geometric Potpourri Seminar 169
Geometry
  analytic 77
  computer 176-178
  descriptive 1,45
  hyperbolic 40,160
  perspective 45-46
  projective 82-83
  synthetic 77
Glanzpunkt 62
Gluing
  diagram 150,*152*,153,*154*,387
  pattern 70-71
  scheme 69
Golden ratio 99,*100*
Gray, Stephen
  hidden line drawing *30*,31
Gradient flow 160
Group
  action 70
  braid 129
  braid, center of 134
  braid, spherical 134,136
  crystallographic 72
  Euclidean congruence 26
  fundamental 71,74,125-126,129,157
  isometry 149
  holonomy 74
  homeotopy, *see* mapping class
  homology 173
  homotopy 125
  Magnus 137-138
  mapping class 126,128-129,136, 144,148
  monodromy 136,173
  orthogonal 59

H

Haken, Wolfgang
  incompressible surfaces 156
Halpern, Benjamin
  perspective 46
  Boy surface node 89
Half-tone shading 63
Harmonic conjugates 48
Height function 148
HNN-extension 72
Heptahedron 86
Holonomy group 74
Homeomorphism 126,143
  see also self-map
Homotopy
  null 125
  product 179
  regular 106
  see also deformation
Hoop
  earphones 107
  double 88
  wobbly 3,83,85
Hurwitz, A.
  monodromy theory 136
Hyperbolic geometry 40,160
Hysteresis 16

I

Idászak, Ray
  rendering algorithm 177-179
Illumination
  experiment *61*,63
  inverse square law 62
  theory 60-61
Illusion
  of curvature 177
  depth 43
  dice 71
  Escher 20
  full moon 62
  Necker cube 20
  Mach line 62
  railway track 54
  optical 18
  perspective 52,*53*,57
  Penrose 68
  tribar 65
  triprong 18
Immersion
  projective plane 89,92
  see also mapping

Implicit function theorem 159
Indicator path 131,139
Interpolation
  linear 5,85
  trigonometric 5,85,117
Isophote 63
Isotopy
  ambient 9
  shear 170
Isys FORTH 177

J

Jacobian determinant 11,86-87,116

K

Keel
  Peano saddle 88
  Whitney umbrella 79
Klein bottle, see surface
Knot
  cable 38-39,40,114
  companion 36,38
  complement 149,151
  diagrams 91,174
  diapered trefoil 27,II
  figure-eight 36,149,*150*
  framed 174
  four-knot 150
  group 28
  projection 28
  spanning surface 37,155,169

L

Lambert, Johann
  cosine law 60,177
  see also shading
Lamination 36,*37*
Latitude of forms 14
Line
  border 7
  construction 16
  contour 2,4,7,8-9,120
  density 14,16
  double 7,92
  drawing 16,*33*,34
  hidden 16,26,31,33
  horizon 46,*49*
  ideal 45
  projectors 43
  sight *44*,46
  visible 16

Line pattern
  blackboard 26,27
  Cayley cusps *10*
  computer based 29,32
  defined 16
  dunce hat *21*,23
  Möbius band 15
  Owl and Pussycat 161
  Romboy homotopy 95
  saddle *2,4*,17
  Seifert surface 155,166
  Whitney umbrella 7,79
Link
  Borromean *42*
  number 161
  vertex 22,*23*
Loop
  bouquet 126
  inverse 125
  hysteresis 16
  product 125

M

Mach line 61-62,64
Magnus, Wilhelm
  monodromy group 136,138
  spherical braid group 134
Manifold, see also space
  acylindrical 39-40
  Euclidean space form 74,*75*
  geometrical 149
  Gieseking 149-150
  Penrose tribar *66*,68
  Riemannian 74
  spine 21
  three dimensional 70
  three-sphere 149
  three-torus 74
Matte surface 63
Mapping
  branched covering 137
  embedding 7,9
  excellent 116,120
  first return 160
  Gauss 87-88,103
  homeomorphism, see self-map
  immersion 7,9
  isotopy 9,143
  Meyer-Vietoris 174
  monodromy 136,160
  proper 8
  pseudo-Anosov 160
  self, see self-map

smooth 8
stable 8
torus 160
Max, Nelson
  eversion film 29
  picture proof 123
Möbius, August
  band, *see* surface
  Morse theory 142,148
Model
  Beltrami-Klein 36
  Poincaré *35*,36
  three dimensional 78
  wire mesh 96,109
Morin, Bernard *see also*
  eversion
  descriptive geometry
  move
  surface
  twist
Move
  Gauss switch 169
  Morin tobacco pouch 122
  Seifert ribbon 161
  Shapiro baseball 122
  turn-reflection 151,172
Murasugi sum 174

N

Neck
  torus 103,*123*
  gastrula *113*,122,*124*
Number
  characteristic 102
  cross caps 102
  double segments 101-102
  of handles 102
  linking 162
  tangent winding 104
  twisting 161
  writhing 161
Neil, William
  semi-cubical parabola 11
Nielsen, Jakob
  automorphism group 128
Node 91-92,178
Normal surface 6,9,117

O

Occlusion 15
Oresme, Nicole
  latitude of forms 14

Orthocenter 49
Orthogonal group 59
Orthonormal matrix 59
Ovalesque 96,III,IV
Overhead projector 28

P

Patch
  regular 6
  see also window
Painting
  pinch point 92
  surfaces 89
Peano saddle *87*,88
Petit, Jean-Pierre
  eversion drawings 109
Perspective
  aerial 43,*61*,64
  affine 43
  architectural 45
  chimera *81*,82
  Chinese 122,*123*
  circle 56
  cube 46,53
  geometry 44-46
  linear 43,*44*,*47*,61,64
  marginal distortion 56
  one-point 52
  panoramic 56
  railway track 54
  scaling parameter 45
  square 53
  three-point 49-50,52,*73*
  two-point 50,*52*
Photography 176
Picture
  element, *see* pixel
  mnemonic 159-10
  pictograph *15*,26-27
  plane 6
  polychrome 27
  solid object 43
  working figure 27
Pill box, *see* coin
Pitchfork bifurcation *10*,11
Pixel
  coordinate 5
  cloud 177,178
  colored 177
Plan and elevation 43
Plate trick 134
Plato's cave 89

Plane
  base 92
  hyperbolic 5,36
  ideal 45-46
  picture 6,46
  projective 36,45,77,89,92
  source 8
  support 86,89,96
Plumbing 107,173
Poles, north and south
  Apery surface 96,177,178,IV
  sphere 104
Poincaré conjecture 22
Point
  base 125
  branch *12*,13
  brilliant 62
  critical 148
  diagonal 54,*55*,78
  double 8,16
  ideal 45,*94*
  pinch 7,8,*12*,92,102,120,137
  recombination 91,95
  regular 6
  singular 8
  triple 8,*90*,91,*94*
  unstable 8
  vanishing 46,*47*,82
  zenith 46,*47*
Porteous, Ian
  plate trick 134,*135*
Principle
  Hessian 86,88-89
  Mach 64
  Thales 52,85,88
Profiling 16,*18*
Projection
  affine 20,43,65
  cabinet 58
  cavalier 58
  isometric 60,*70*
  oblique *15*,43,58,60
  orthographic 43,54,*58*,59,65,*73*
  parallel 43
  perspective 43
Projectors 43
Pugh, Charles
  eversion models 109

Q

Quaternions 134

R

Rainich, Yuri
  Greek geometry 26
Regular homotopy 105
Reference frame *2*,*7*,14-15,26
Rendering
  computer 177-178,III,IV
  errors 177
Ribbon
  curling 161
  twisting 119,155,161
  writhing 161
  neighborhood 143
Roads 153
Roll
  matrix 146
  self-map 146
  Solomon seal 147
Rolling cusps 91
Romboy homotopy
  Apery 96
  Idaszak-Cox 178-179
  Morin 95

S

Schematic diagram 149
Sauze, Max
  wire mesh models 96
Seam 151,157
Self-map
  coil 134
  commuting 128
  handle 139,*140*,141
  identity 126
  product 126
  punctured plane 134
  punctured sphere 134
  roll 146-148
  superposition *127*,128,*141*
  surface 126
  swap 128
  twist 128,144
  whorl 138
Shading
  cusps 17
  folds 64
  half-tone 63
  guides 18
  Lambert 62,177
  quarter-tone 63
  shadows 61

INDEX

Shapiro, Arnold
  sphere eversion 104,106
Shear isotopy 170,173,174
Sight cone 48
Singularity
  elliptic umbilic 13
  fold type 9
  point 8
  swallowtail 17
  theory 9
Slides, 35mm 27-28
Sliding coefficients
  Romboy homotopy 96,178
  trigonometric 85,117
Slit
  branched covering 138
  Riemann surface 137
  complementary 137
  surface complex 151
Smale, Stephen
  eversion theorem 105,122
Solid geometry modeler 176
Space, *see also* manifold
  projective 82-83
  target 8
Specular reflection 177
Spine of a 3-ball 21
Stable
  approximation 9
  forms 1
  element 72
Stalk 80
Stacking
  Boy surface *90*,91
  chapeau 108
  cross cap 95
  Whitney bottle *94*,95
Steiner, Jacob
  synthetic geometer 77
Stereograph 31
Stereoscope 31
Stillwell, John
  curls 16-17
  pipeline 68,*69*
Style,*see* drawing style
Suiseth, Richard
  latitude of forms 14
Surface
  biting discs 151,*152*,157
  Boy 17,*90*,101,105,114,172,178,IV
  Cayley cusp *10*,11,24
  chapeau 104,*108*,112
  chimney *137*,138

cross cap *87*,88-89,102,177-178,IV
cubic *10*,11
curved 78
double torus 126,*127*,138,164,*165*,166
doubly ruled *2*,3
dunce hat 20,*21,23*
earphones *105*,107
embedded, *see* mapping
Enneper *118*,119
Etruscan III,IV
Euler characteristic 102
excellent, *see* normal
general position, *see* normal
genus 102
Haken 156,*158*
Ida 179,IV
immersed, *see* mapping
heptahedron 86
hyperbolic paraboloid 2,*4*
Klein bottle 17,*118*,119,157,178,III,IV
knot spanning *37*,155
matte 63
minimal 24,119
Möbius band *23*,24,*25*,
  27,45,*101*,102,*115*,143,157
Möbius disc *15*,80-81,87-88,95
Morin *17*,29,110,*113*,117
Morin band 119,122
singular Möbius band, *see*
  Möbius disc
normal 6,9,117
owl 161
ovalesque 96,198,IV
pair of pants *34*,36
painted 92
Peano saddle 88
Plücker conoid *81*,82
pussycat 161
polyhedral 17,*19*,20
Riemann 13,102
Roman 17,77,*84*,83-86,96,178,IV
ruled 3,116
saddle 11,20,85
Seifert 107,*110*,155,*159*,157,160
self-map 126
singular torus 148
Solomon Seal *145*,147
stable, *see* normal
Sudanese Möbius band 24,*25*
torus *159*
trigonometric *30*,31,*103*
Veronese 83, 88
Whitney bottle *94*,95
Whitney umbrella 7,8,*81*,148

Swallowtails 17,114
Swap
  braid 129,131,*132*
  handle cores 142
  matrix 146
  relator 129
  self-map 144,148
Switching operation 169,173
Symmetric torus 147

T

Tait, P.G.
  sailor's knot 50
Tangent winding number 105
Tees 153
Thales principle 52-53,85,88
Thom, René
  dialectic models 14
Thurston, William
  grand drama 166
  knot spanners 28
  tripus 153,155
Tobacco pouch *115*
  eversion 114
  move 122
Topology
  classification 117
  descriptive 1,12,45,117
  Whitney 8
Topological equivalence 102
Transitional form
  double torus knot 165-169
  plastic models 86
  shear isotopy 174
Transparencies 27
Triprong 18
Twist
  Dehn 122-123,138,142,174
  half 138
  Lickorish 142
  Morin 120,123
  ribbon 68,161,173
  self-map 143,146
  Turn-reflection 157
Tunnels 155

U

Umschaltung 169
Undercut 64
Underlay 31,36,40
UNIX 177

V

Venus de Iris 179,IV
Veronese, Guiseppe
  analytic geometer 77
Vertex link *22*
Videotape 176
Visual cone 46

W

Window
  chalk drawing 26,*I*
  regular homotopy 112,*113*
  surface patch 6,7
  visualization aid *111*,162
  Whitney bottle 95
Wire frame 31,*32*
Whisker 34,80
Whorl
  braid 131,137
  matrix 146
  relator 129
  self-map 138
Wobbly hoop 3,85
Working figure 27
Worm holes 153
Writhing
  number 161
  ribbon 68

Z

Zeeman, Christopher
  catastrophe machine 120
  conjecture 22
  dialectic model 14,16
Zones
  Boy surface 108
  Cayley cusp *90*
  Whitney bottle *94*,95